Astronomers' Univer

For further volumes:
http://www.springer.com/series/6960

Steve Miller

The Chemical Cosmos

A Guided Tour

 Springer

Steve Miller
Department of Science and Technology Studies
University College London
Gower Street, WC1E 6BT London, UK
s.miller@ucl.ac.uk

ISBN 978-1-4419-8443-2   e-ISBN 978-1-4419-8444-9
DOI 10.1007/978-1-4419-8444-9
Springer New York Dordrecht Heidelberg London

Library of Congress Control Number: 2011937447

© Springer Science+Business Media, LLC 2012
All rights reserved. This work may not be translated or copied in whole or in part without the written permission of the publisher (Springer Science+Business Media, LLC, 233 Spring Street, New York, NY 10013, USA), except for brief excerpts in connection with reviews or scholarly analysis. Use in connection with any form of information storage and retrieval, electronic adaptation, computer software, or by similar or dissimilar methodology now known or hereafter developed is forbidden.
The use in this publication of trade names, trademarks, service marks, and similar terms, even if they are not identified as such, is not to be taken as an expression of opinion as to whether or not they are subject to proprietary rights.

Printed on acid-free paper

Springer is part of Springer Science+Business Media (www.springer.com)

For Vanessa

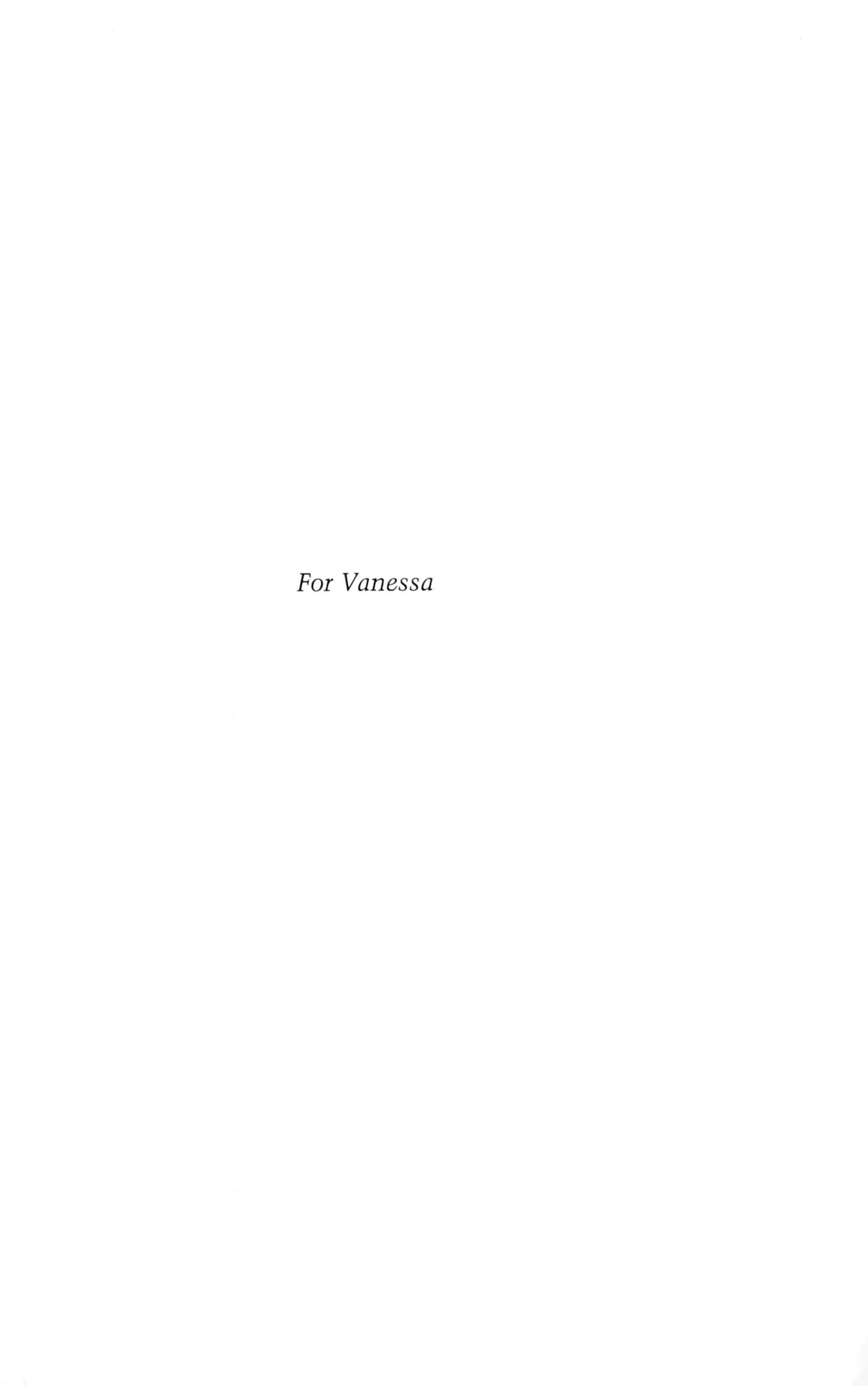

Acknowledgements

This book was largely written whilst I was on sabbatical leave from University College London (UCL) in 2009 at the Institute for Astronomy (IfA) in Hilo, Hawaii. So I would like to thank my Dean at UCL, Professor Richard Catlow, and Professor Alan Tokunaga, Director of the NASA Infrared Telescope Facility and my host at the IfA. Professor Bob Joseph, also of the IfA, introduced me to Hawaii and infrared astronomical observing, and shared much of his great enthusiasm for both with me. Over my 25 years at UCL, it has been an enormous pleasure to work with some great friends and colleagues in both the Department of Physics and Astronomy and the Department of Science and Technology Studies, and their support and encouragement in my various enterprises is much appreciated. Professor David Williams (UCL), Dr Tom Stallard (University of Leicester) and Dr Declan Fahy (American University, Washington) all read various versions of the book, and their insightful and helpful comments have improved it enormously. (The faults remain mine, however.) I would like to thank the editorial team at Springer – Jessica Fricchione and Harry Blom – for their advice and patience. Above all, this book has been inspired by the work of Professor Jonathan Tennyson (UCL) and Professor Takeshi Oka (University of Chicago). Long may it continue.

Contents

	Acknowledgements	vii
1.	Purple Haze: Introducing Our Guide	1
2.	The Early Universe: The Source of Chemistry – and of Our Guide	9
3.	Shooting the Rapids: The Life and Death of the Earliest Stars	25
4.	Heading Downstream and Cooking by Starlight	63
5.	Fishing for Molecules	91
6.	Branching Out: In the Land of the Giants and Dwarves	115
7.	In the Delta: Exoplanets – Worlds, but Not as We Know Them	153
8.	Towards the Sea of Life	171
	Epilogue	191
	Annotated References and Further Reading to Chapters	195
	Some Useful Numbers	221
	Pictures and Figures	223
	Index	227

Prologue

In the beginning, there was Hydrogen. And not a lot else. Okay, there was some Helium, Lithium and a heavy form of Hydrogen called Deuterium. But there was none of the Carbon, Oxygen, Nitrogen, Sulfur, Phosphorus, Calcium, Sodium, etc. that are vital to our very existence. But here we are, and today we know of 110 chemical elements forming literally billions of chemical compounds. Some of these compounds are sufficiently ingenious that they can replicate by themselves; some of them are sufficiently sociable that they team up to form living creatures – algae, bacteria and – eventually – life-forms such as ourselves. So how do we get from Hydrogen (plus a few friends) to where we are now? The answer is provided by astronomy, the study of the heavens bright and dark.

Astronomy is a journey: it is a journey over enormous distances to other worlds, other stars and other galaxies. It is also a journey back in time. Light takes time to cross the vast distances of empty space. So astronomers are always looking at other worlds, stars or galaxies as they were when the light by which we see them first left home to reach us. In this book, we shall take a *chemical* journey, following the flow of the Chemical Cosmos from its source in the early universe all the way down to the sea of life. So vast is the journey that we will need a guide, one with an adventurous spirit, one prepared to endure many hardships, and one that will pop up when we most need it, but least expect it. Our guide will be of simple but ubiquitous parentage. It will be both stable and energetic; it will have been there since the beginning of the Chemical Cosmos, and it will be there at its end.

Some time before the end of the decade, or thereabouts, if enough money can be found, a huge space telescope will blast off from a launch site in French Guyana. The James Webb Space Telescope will be ten times as powerful as the current Hubble Space Telescope. It will examine the sky in the infrared part of the spectrum – wavelengths longer than visible red light, responsible both for heating and for cooling the universe. What it will probe is the Chemical Cosmos, the river of astronomical chemistry that has its source in the early universe and takes us all the way to the sea of life. Much of what the James Webb Space Telescope finds will be due, directly or indirectly, to our guide along this river journey. Our guide needs an introduction.

1. Purple Haze: Introducing Our Guide

Outside of Chicago's City Hall is a giant Picasso sculpture of a weeping woman. For the more artistically challenged, it takes quite a while before you can "see" it, before you can really make out what Picasso was getting at and how he got there. Five miles to the south of City Hall, in the basement of the University of Chicago's Chemistry Department, lies a piece of glassware of which the great artist would have been proud.

Again to the uninitiated, it takes quite a while to "see" it. It looks like a deranged spider; indeed, those who work with it call it the Tarantula. When it is working in the darkened laboratory in which it sits, it is suffused by a purple haze and resonates to an electric hum. The Tarantula is not a work of art in the conventional sense, although it is certainly a tribute to the art of the glassblower who made it. This artistic glassware is a discharge tube, a device for making electrically charged chemicals that are normally only found high up in the atmosphere or in the depths of outer space.

We will be returning to the Tarantula shortly.

The Tarantula's owner is Takeshi (just call me) Oka, (now Emeritus) Professor of Chemistry and Astronomy, graduate of the University of Tokyo, distinguished member of the British and the Canadian Royal Societies, holder of many other distinctions from a scientific career that now spans six decades (Figure 1.1). In Chicago, Oka runs the *"Oka Ion Factory"*, a laboratory that has paved the way in the study of chemicals that are called "molecular ions".

Ions derive their name from the Greek *ion*, meaning "moving thing," and they were given this name by Michael Faraday, Professor of Chemistry at the Royal Institution in London between the years of 1833 and his death in 1867. Ions, explained Faraday, are what move in a chemical solution, or – in a more modern

2   The Chemical Cosmos

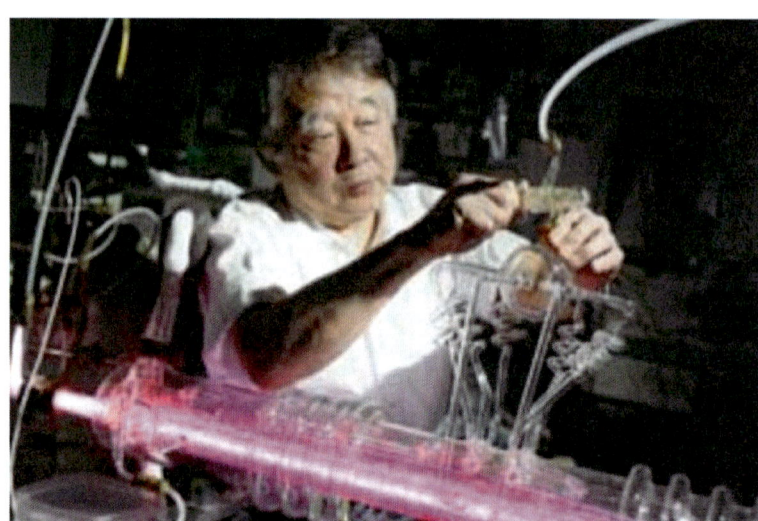

FIGURE 1.1 Takeshi Oka at work in his laboratory at the University of Chicago: *credit – Oka Ion Factory, University of Chicago.*

application – a fluorescent light tube, when you run an electric current through it. Opposites attract – *cations* are positively charged, and travel towards the negatively charged cathode. Conversely *anions* are negatively charged and head for the – you guessed it – *positively* charged anode.

The smallest element of negative charge is called the electron, the first sub-atomic particle ever discovered in 1897 by the British physicist Joseph John (J.J.) Thomson (Figure 1.2). Atoms are made up of electrons surrounding a nucleus, positively charged protons and electrically neutral neutrons. Atoms may become positively charged by dumping a negatively charged electron; and they then become cations like the Sodium atom in common table salt. Or atoms may become negatively charged by picking up an electron and then become anions like the Chlorine atom in the same salt crystal.

Molecules are groups of atoms more or less tightly held together, like Water. In Water, two Hydrogen atoms combine with one Oxygen atom to form the Water molecule. Molecular ions are electrically charged molecules that have either been careless with their electrons – molecular cations – or greedy for them – molecular

FIGURE 1.2 J.J. Thomson giving a lecture demonstration in the Cavendish Laboratory at the University of Cambridge: *credit – The Cavendish Laboratory, University of Cambridge.*

anions. Molecular ions are literally everywhere, and even in Water, that benign prerequisite of life as we know it, one molecule in ten million has had enough of neutrality and become a cation. And, to maintain electrical balance, one has become an anion; therefore scientists find molecular ions fascinating.

Oka with his Ion Factory is to molecular ions what Henry Ford was to automobiles. (The Ion Factory could also have been called the Professor Factory; there is many a university around the world who owes at least one of its Chemistry professors to the training they received at the hands of Oka, and fellowship his lab generated.) But this is not the story of the Oka Ion Factory itself, although we shall return to it again in our story. Our adventure goes way beyond the confines of the University of Chicago, far out into space beyond our galaxy, the Milky Way, and far back in time to an era in which very few of the chemicals that make up our world had been formed. On our adventure, we shall follow the fortunes of a tiny triangular adventurer, so small that ten billion of them standing in line stretch for little more than a meter.

Our guide is a molecular ion that goes by the name of H_3^+ (read H-three-plus, if you want to). So what, exactly, is H_3^+ you may ask?

For starters there is a big clue or two in the name. All elements have a chemical sign to indicate their atoms – H for Hydrogen, He for Helium, C for Carbon, N for Nitrogen, O for Oxygen, Cl for Chlorine, etc. So you can see that the chemical signs are either one or two letters long. When atoms combine to form a molecule, the molecule gets its own chemical symbol, known as a formula, derived from the atoms that make it up. The formula for common salt is NaCl, which shows that it is made up of equal numbers of Sodium (Na for the Latin word, Natrium) and Chlorine (Cl) atoms. The formula of Water, H_2O, indicates Hydrogen atoms combining with Oxygen in the ratio of two-to-one.

But H_3^+ only has 'H' in it; there are no other atoms in it. In an exclusive fashion, in H_3^+, Hydrogen has decided simply to combine with itself, and it turns out that this is not so unusual. Oxygen atoms like to hang around in pairs, if there is no better offer at hand, to form O_2 molecules, the stuff of air that we take in through the walls of our lungs to keep us alive. Nitrogen and Chlorine atoms will also happily keep each other company, as N_2 and Cl_2. And Hydrogen is most often found doubled up as the molecule H_2.

Nor is three necessarily a crowd. Oxygen atoms will hold hands with two others to form Ozone, O_3, a pollutant at street level but a life saver high in the Earth's atmosphere where it blocks out harmful ultraviolet radiation. Indeed, if it were not for Oxygen "tripling up" in the form of Ozone, life on Earth would be impossible today. And there are many atoms that will form huge conglomerates. Pure Carbon is the most prolific of them all; it forms endless chains in graphite, extensive crystals in diamond and ball-shaped clusters of C_{60} – 60 atoms of Carbon joined together in the form of a miniature soccer ball – and even bigger.

So we should not be surprised at three Hydrogens hanging out together, although – as we will see later – it actually was a surprise when it was first discovered.

The second clue from the name is the plus sign – H_3^+. This means that we are being introduced to a cation, positively charged. In former and more formal times, when someone was introduced, it was customary to enquire after the family to which the newcomer belonged. After all, one did not want to be consorting with any old riff-raff, one wanted to be sure that one was talking to the *right* Kennedys or the *right* Windsors.

Chemists tend to think of ions as being the offspring of parent atoms or molecules. In common salt, Sodium exists as a positive cation – Na^+ - and Chlorine as a negative anion – Cl^-. These ions are the children of their neutral parents, Na and Cl respectively. Protocol has been observed; we are talking to the *right* Sodium cation and the *right* Chlorine anion. Sodium is a great guy, stable and well respected, part of the Alkali Metal clan whose ancestors go all the way back to the Big Bang. And you could not wish for a nicer girl than Chlorine, a member of the bustling Halogen family. No wonder they have such great ions as offspring and that those offspring go so well together.

Any logically thinking person by now would have worked out that the parent of H_3^+ is good old H_3. But H_3 is the parent you do not really want to talk about: H_3 is unstable and as elusive as an "ex" behind with the alimony, bringing us back to the Tarantula in the Oka Ion Factory, glowing purple as the electricity flows through it.

Hydrogen is the simplest of all atoms made up of a nucleus that is a single, positively charged proton. All atoms have protons, but Hydrogen has only one. The proton is over 1,800 times more massive that the electron, and the positive charge of the proton is balanced by the negative charge of just one electron – together they make up the Hydrogen atom. This means that the Hydrogen ion, H^+, and the proton are one and the same.

Oka's Tarantula can be filled with pure Hydrogen gas, the paired up H_2 form. As the electricity flows through the gas, some of it is ionized – broken up into loose electrons and positively charged Hydrogen ions, swimming in a sea of ordinary Hydrogen gas. Although the gas is at low pressures, Hydrogen molecules and Hydrogen ions bang into each other millions of times every second, sometimes sticking together. The net result of all this excitement is that a neutral Hydrogen molecule, H_2, picks up a Hydrogen ion, H^+, to form our adventurer H_3^+.

This process turns out to be one of the most fundamental of all chemical reactions in the universe. We encounter it not just in the basement of the University of Chicago's Chemistry Department, but in the atmospheres of the giant planets like Jupiter and Saturn. We also encounter it in other planets that exist beyond our Solar System, in the top layers of stars that are among the earliest ever born, and in the vast gas clouds that fill up not just the Milky Way, but galaxies as far as we can see. It is a process that is nearly as old as the universe itself, much older than the formation of Water or common salt.

So now when H_3^+ introduces itself, it can keep quiet about its wayward parent H_3. Instead, it can boast of the proud union between the stable and respected H_2, a molecule with quite literally 'universal' appeal, and the most fundamental of all nuclear particles, the proton H^+. Indeed, our chemical guide can say, "I'm *Protonated* Hydrogen."

Hydrogen is the most abundant chemical in the universe; nine out of ten atoms are Hydrogen. Helium makes up almost all of the rest, and the Carbon, Oxygen, Nitrogen, and all the other atoms that are so important for the framework of our Earth and ourselves add up to only one thousandth of the atoms in the universe. So any molecule that can boast a parentage of pure Hydrogen is part of a very prolific tribe.

It turns out that, unlike the wayward parent H_3, the offspring H_3^+ is a stable chemical and its parts are strongly bound together. It can boldly go into some of the most challenging of environments, but because it is an ion – a positively charged cation – it is very reactive. So it makes things happen everywhere it goes. Child of the most abundant species in the universe, reactive in a way that none of its relatives can match – that is why the adventures of H_3^+ are the most energetic and far-reaching we can wish for.

Our adventure with H_3^+ will take us to the giant planets Jupiter, Saturn and Uranus. It will take us out of our Solar System and into the planetary systems that have been discovered around nearby stars, stars that are to be found within a few to a few tens of light years from the Sun, but which are probably typical of billions of billions of stars within our own galaxy, the Milky Way, and galaxies that lie beyond it. Voyaging with our little chemical guide, we shall traverse giant clouds of gas and dust that lie between the stars all the way to the center of the Milky Way, where a giant black hole consumes all who venture too closely. On the way we will visit some of the oldest stars in the galaxy and will even journey to our neighbor galaxy, the Large Magellanic Cloud, to see what H_3^+ can tell us about the death of a star many times larger than our own "little" Sun. And it may be that our adventurer played a part in ensuring that the Solar System evolved in such a way that life on Earth could evolve.

Our guide also takes us into a world little appreciated here on Earth, although ubiquitous in space – the world of plasmas. The Greek philosopher Aristotle had four elements – Earth, Water, Air, and Fire. Today, we consider everyday matter to exist in the states of solid, liquid and gas, which could be thought of as corresponding to Aristotle's Earth, Water and Air. Plasma is the fourth state of matter, not fire; plasma consists of electrically charged gases of very low density. The Solar System is filled with plasma: the solar wind, a stream of electrically charged particles that pour out continuously from our Sun causing beautiful aurorae and destructive electrical storms, is plasma. Plasma is the home of H_3^+, and it is in this environment that our guide first lights the chemical fires that lead all the way to the building block of life itself.

First, though, we will journey back in time some 13½ billion years to start of our universe, the source of our river of cosmic chemistry. Way to go for such a simple little molecule!

2. The Early Universe: The Source of Chemistry – and of Our Guide

On March 30, 2010, an experiment called the Large Hadron Collider (LHC) succeeded in crashing together two beams of protons at the colossal energy of 7 million million electron volts. (An electron volt is the energy given to one electron passing through an electric field of 1 V.) This was energy 3½ times greater than anything achieved before, and made up for a nervous 18 months while scientists waited to see if the billions spent on the LHC were justified. This enormous particle collider is housed in a vast tunnel spanning the border between France and Switzerland at the European Nuclear Research Centre (CERN) near Geneva. Operating 100 m underground, the LHC is the latest in a long line of experiments designed to investigate the world at a sub-atomic level and is now the most powerful tool at the disposal of scientists who work in the area of particle physics. With it, particle physicists are attempting to recreate the conditions of the very early universe.

Immediately after its birth – at least, if the current theories are to be believed – the universe was a very energetic place. Protons and electrons ran around freely, along with neutrons – neutral particles with a mass very similar to the proton – while a zoo of other more exotic fundamental particles rushed to and fro like traders in a bear market. In addition to the particles of matter, there were also the particles of light known as photons, particles that have no mass of their own, and because the negatively charged electrons and positively charged protons interact strongly with light, photons were "trapped" in with the ordinary matter in a hot, vigorous soup.

This brief sketch – and it is just that – derives from the best theory that we currently have to explain the universe that we live in. Because it starts with an "explosion" of truly cosmic proportions, it was nicknamed the "Big Bang" by people who did not believe in it, and who began to ridicule it. The Big Bang Universe is not just a whim, though, because it is strongly supported by scientific evidence – the expansion of the universe measured by galaxies and clusters of galaxies racing away from one another, the discovery of the afterglow of the initial explosion, and, crucially for our story, the chemical composition of the universe. Indeed, the Big Bang was initially proposed to explain the whole of cosmic chemistry.

The biologist J.B.S. Haldane was once asked if he could deduce anything about God from his study of the natural world. So the story goes, Haldane replied that if He did exist, the Creator had "an inordinate fondness for beetles" – they are everywhere, in species too numerous to name. Astronomers who were asked the same question might answer to the effect that God had "an inordinate fondness for Hydrogen". Hydrogen is the lightest and simplest of all atoms, comprised of just one positively charged proton orbited by one negatively charged electron. It, too, is everywhere; some nine out of ten of all atoms are Hydrogen atoms, and it makes up nearly three fourth of the mass of ordinary matter in the universe.

Although Hydrogen is the lightest and the most abundant of all elements, it is not alone in the universe, which is fortunate for Carbon-based life forms such as ourselves. It is joined by a 100+ series of heavier elements, the next heaviest and most abundant element being Helium, Element 2, which makes up 24% by mass of the ordinary matter of the universe. Carbon, Element 6 and 12 times as heavy as Hydrogen, makes up just half a percent of the ordinary matter mass; Oxygen, Element 8 and 16 times heavier than Hydrogen, makes up just 1%. In between them, Element 7, Nitrogen, contributes just a tenth of a percent to the mass of ordinary matter. As the element number and the mass increases, so the proportion found in the universe decreases, at least until the very heavy elements are reached.

In the immediate aftermath of World War II, with the images of the atomic explosions of Hiroshima and Nagasaki still fresh, George Gamow of George Washington University pointed out that one could explain the fact that there were fewer heavy chemical elements than light ones if the early universe were in a highly *unequilibrium* state – far out of energetic balance with itself – and was expanding and cooling rapidly following an initial explosive event. Since the nucleus of heavy elements would take longer to build out of the fundamental protons and neutrons that made it up, heavy elements would be rare if the time available to make them were short. And time *was* short for the expanding universe, product of the Big Bang explosion, was both rapidly cooling and getting less dense. So the chances of sufficient protons and neutrons coming together with enough energy to produce heavier and heavier elements got slimmer and slimmer as time went on. This was why, Gamow argued, the abundance of heavy elements would fall off dramatically as the element became heavier – which was exactly what astronomers observed as well.

Gamow was right but the trouble was he was too right. Calculations on the Big Bang universe showed that the temperature and density of the early universe fell so rapidly that all that could be formed were the nuclei of Elements 1 through 3 – Hydrogen, Helium and Lithium – and Deuterium, a heavy form of Hydrogen that we will come across later. That made the early universe chemically simple, with just three chemical elements, but left unanswered how heavier elements, such as Carbon, Nitrogen and Oxygen and the other 100-plus elements, were formed. To answer that question, a subtle blend of astro-physics and astro-chemistry is required.

The early universe is clearly a product of what happens in the physical Big Bang and its immediate aftermath. It did not really start to be a *chemical* universe, though, until at least a hundred thousand years after the initial cosmic explosion, and probably more like 300,000 to 400,000 years. By that time, the temperature of the universe had fallen to a "mere" 4,000 degrees cooling to 3,000 degrees above absolute zero, still hot enough to melt almost anything except diamond (not that diamonds existed at this time, since there was no Carbon to form them) but cool compared with earlier times. (From now on, we will use the symbol K to denote "degrees above absolute zero". K stands for kelvin, and absolute zero is –273.15 degrees Centigrade. Note that a kelvin is the same temperature interval as a degree Centigrade. So temperatures expressed in kelvin, K, will always be 273.15 greater than temperatures expressed in degrees Centigrade.) Once the temperature got below about 3,000K, about 300,000 years after the Big Bang, positively charged protons – the nucleus of a Hydrogen atom – could (re-)combine with negatively charged electrons to form neutral Hydrogen atoms for the first time; electrons had teamed up with Lithium and Helium nuclei to form neutral atoms a "bit" earlier. Meanwhile, photons, the particles of light that had been trapped in the proton/electron soup of the very early universe, could escape and wander free. Matter and radiation were now no longer tightly coupled and could act independently of one another.

This period, extending from around 100,000 to 400,000 years after the Big Bang, is called the "Recombination Era," and more than 13 billion years later, we can still measure the light that first escaped at the time of the Recombination Era. Over time, and as the universe has expanded, the temperature of this all-pervasive background radiation has cooled from some 3,000K to just 2.73K and its wavelength has lengthened from (infra-)red to microwave, but it is there wherever we look out into space. This Cosmic Microwave Background Radiation, as it is known, was discovered in 1964 by American radio astronomers Arno Penzias and Robert Wilson, and for most astronomers that pretty much ended the argument about whether the universe started with a Big Bang or whether it had existed in a steady state from time immemorial.

So the Cosmic Microwave Background Radiation is the oldest radiation in the universe, and it carries in it the imprint of what the cosmos looked like and how it was structured in those early days. As well as allowing us to understand the universe at early times, however, it also acts as a veil; although we can derive theories about what went on before, and even try to simulate what happened in enormous particle accelerators such as the LHC, we cannot actually see further back in time than the time at which recombination happened, the time at which atoms started to form. The first few hundred thousand years of the universe are veiled off from direct observation, no matter how powerful our telescopes or how sensitive our instruments.

The Recombination Era produced the first electrically neutral atoms in the universe. It sounds easy enough; opposites attract and an electrically positive atomic nucleus and one or more negative electrons team up to form a neutral atom. The problem, however, is *excess* energy since free electrons and free atomic nuclei whizzing around the universe have enough energy to keep each other at arm's length. If they are to team up to form a neutral atom that is stable, they cannot keep all that energy; if they do, they will simply fly apart again. After all, in any relationship there has to be a bit of softening, a bit of accommodating to the partner's needs if things are going to work out. Just how depends on the fundamental structure of the atom itself.

The notion that matter consists of atoms – literally "uncuttable" – goes back at least to the Greek philosophers Leucippus and Democritus, who lived in the fifth century BC. According to these philosophers, the properties of materials could be deduced from the properties of the atoms from which they were made. Atomic theory began to take on its modern form with the work of the nineteenth century Manchester chemist John Dalton, whose ideas included the notions that the atoms of any particular chemical element were identical, and that chemical reactions involved the rearrangement of atoms but could neither create nor destroy them. Once the nuclear reactions in the immediate aftermath of the Big Bang had ended, chemical reactions in the early universe might rearrange the atoms that had been produced but could not change the overall composition of elements. For some chemists like Dalton atoms were real; for others, however, they remained merely a "convenience", a way of keeping the chemical books straight whilst following the ever more complex reactions and sophisticated compounds that nineteenth century chemistry involved.

The year 1905 was a marvelous year for a young patent clerk called Albert Einstein, a man who would turn out to be one of the greatest minds of the twentieth century. It is best remembered as the year that he put forward his theory of special relativity, commencing what the London *Times* would later call a "revolution in science" that "overthrew" the classical mechanics of Sir Isaac Newton. (Einstein himself was far more modest in describing his achievements.) Less appreciated, however, is the work he did on what was called "Brownian motion".

Brownian motion was probably first described in writing in 60 BC by the Roman poet Lucretius in his *De rerum natura* (On the nature of things). Lucretius described the random "dancing" of particles of dust caught in a beam of sunlight as being due to "underlying movements of matter that are hidden from our sight" caused by the impetus of atoms, an idea that he inherited from Leucippus and Democritus. However, Brownian motion is actually named for the botanist Robert Brown who observed the same random dancing of pollen grains in water. What Einstein did was to show mathematically that the intuition of Lucretius was right, giving conclusive proof to chemists that the atoms that they had *proposed* as a chemical convenience really did exist.

The first real understanding of the structure of the atom is due to the New Zealand-born physicist, Ernest Rutherford. In the early 1900s, Rutherford and his colleagues were studying the newly discovered phenomenon of radioactivity in which atoms, such as Uranium that are unstable, break down and release a variety of "rays". These rays were labeled by the first three letters of the Greek alphabet, alpha, beta and gamma. The beta rays were negatively charged and quickly identified as electrons, themselves newly discovered in 1897 by Rutherford's mentor J.J. Thomson. Gamma rays had no electrical charge and were seen to be very energetic rays of the same sort as light – electromagnetic rays. So what were the alphas? Rutherford showed that they were Helium atoms that had lost their electrons. And he soon showed that these alpha particles could be used to probe the deepest structure of atoms.

Rutherford worked with his assistant Hans Geiger (for whom the Geiger counter that measures radioactivity is named) and student Ernest Marsden to measure the effect of firing a beam of alpha particles at very thin films of metal. Gold was the most suitable because it is very easily worked, and it is possible to produce thin films of gold that are only four millionths of a centimeter thick. Rutherford's expectation was that nearly all of the alpha particles, which were very energetic, would pass straight through the gold foil, but he and Geiger had already noticed that the image produced by alpha particles on a fluorescent screen became "fuzzy" after passing through even the finest of gold films. Clearly the alpha particles were not all passing straight through the film, but some were being deflected off course – by how much and how often though? Marsden was given the task of seeing if any particles were reflected back off the gold film: and there were!

About one in 20,000 or so of the alphas came back off the thin gold film at Marsden along the direction he had originally fired them. Rutherford was astonished: "It was almost as incredible as if you had fired a 15-inch shell at a piece of tissue paper and it came back and hit you" he is reported to have said. It meant that the gold foil was not made up of evenly spread matter, but was a network of tiny, dense obstacles surrounded by almost empty space. Rutherford had put together the basic structure of the atom – a tiny, dense, positively-charged nucleus, surrounded by a space filled with Thomson's negative electrons. How, then, did the electrons "fill that space"?

Spectroscopy was a second tool to probe the structure of the atom. Spectroscopy is to chemistry what fingerprinting is to criminology. Spectroscopy tells you what it is that you have in your test-tube and can claim the highest of scientific devotees - in the 1670s, Sir Isaac Newton used a glass prism to split up white light into its component rainbow of colors, eventually publishing his results many years later in his 1704 book *Opticks*. By the early 1800s, scientists had noted that while sunlight spanned the full spectrum of the rainbow, there were a number of dark lines or gaps that could be seen when precision instruments, much more sensitive than Newton's prism, were used. From 1814 onwards, the German physicist Joseph von Fraunhofer mapped nearly 600 such lines at different frequencies (or colors) in the Sun's spectrum. Many of the individual Fraunhofer lines, as they become known, were later shown to correspond to individual chemical elements, and a line in the red region with a wavelength of 656.3 nanometers (a nanometer, nm, is one billionth of a meter) was produced by Hydrogen. Hydrogen also produced lines in the blue-to-violet region of the spectrum, at 486.1 and 434.0 nm. Sodium produced two lines in the orange very close together, at 589.0 and 589.6 nm. Close by at 587.6 nm was a line that led to the identification of Helium, called so because it was first discovered in the Sun, or Helios in Greek. Some elements were extremely prolific such as Iron which was associated with ten strong Fraunhofer lines from the yellow-green through to the violet spectral regions.

As the nineteenth century closed, one of the major "revolutions" in our understanding of the physical world occurred. German physicist Max Planck proposed that energy could only come in discrete packets, called quanta. Unlike a dollar, which can be used to buy something for 27 cents and get you 73 cents back, quanta do not give you change. It is a quantum or nothing – a bit like a farmers' market where home-grown produce comes in one dollar packs, take it or leave it. Energy does not come in dollars, however, but in packets that are given by the frequency of the light corpuscles – photons – multiplied by a universal constant named for Planck, and given the symbol h. Again in his *annus mirabilis* of 1905, Einstein demonstrated that these packets of energy were real, and that light, which had the properties of a wave, was also composed of particles – again, photons.

Following Rutherford, the atom could then be described as a positively charged nucleus surrounded by "orbiting" negatively charged electrons. However traditional theory predicted that an electron in continuous motion about the nucleus of an atom would radiate away its energy and gradually spiral in until the two hit each other. Danish scientist Niels Bohr took Rutherford's atom together with Planck's quantum theory and simply proposed that this "spiraling in" would not happen if the electron were in an orbit around the nucleus with its angular momentum quantized. For a stable orbit, this angular momentum – given by the mass of the electron multiplied by the speed at which it orbited and its distance from the nucleus – should be a precise multiple of Planck's constant for the quantum of energy, h, divided by two times π, or pi; pi is given by dividing the diameter of any circle into its circumference and has a value of roughly 3.142.

As well as being the most abundant, Hydrogen is also the simplest of all atoms. Its nucleus is a single proton, and this is surrounded by a Rutherford "cloud" of just one electron. According to Bohr's model, the energy of stable electron orbits for Hydrogen would be given by a simple formula that depended simply on the level number, n, multiplied by itself to give n^2. This n^2 was then divided into Planck's constant, h, multiplied by the speed of light, c, and another fundamental constant, R, to get the energy of the level. R was a number known as the Rydberg Constant, and has a value of nearly 11 million inverse meters. The level number n was simply a number ranging from 1, 2, 3 … to as large as you like. The energy was measured from the point at which the Hydrogen atom would break up, or ionize, into a proton to become an H^+ cation, and a free electron. So the formula for the energy of Level n could be written simply as $-hcR/n^2$; the most stable orbits were furthest below the top of this energy "well", hence the minus sign in the formula.

The first energy level was produced when n was 1; in units of hc, it was $-R$ units from the ionization point. (From now on, we will take the hc unit as a given.) The second level was at $-R$ divided by two times 2, that is at $-1/4$ R. A spectral line of Hydrogen due to the electron "falling" from Level 2 to Level 1 has an energy of ¾R, in units of hc, and a wavelength given by 1 divided by that value, that is 4/3R. (This is why the Rydberg Constant is so useful; it leads directly to the wavelengths of Hydrogen lines.) This two-to-one line is actually measured in the ultraviolet part of the spectrum with a wavelength of 121.6 nm and is known as the Lyman-alpha line. The line of Hydrogen seen in the red part of the spectrum by Fraunhofer, known simply as H-alpha, corresponds to the electron changing its orbit from Level 3 to Level 2.

The energy of this line is given, once more, by the difference in energy between Level 3 and Level 2. As we have seen, Level 2 has an energy of 1 divided by two times 2, or ¼ R; Level 3 has an energy of 1 divided three times 3, or 1/9 R. So the energy of this line, again in units of hc is 1/4 minus 1/9 R, or 5/36R. This is equivalent to a wavelength of 656.3 nm. Spectral lines due to electrons in atomic Hydrogen changing their orbit occur right throughout the electromagnetic spectrum. For example, in the infrared region the line corresponding to a change from Level 5 to Level 4, and called Brackett-alpha, occurs at 4,053 nm with an energy equivalent of just a two-and-a-quarter percent of the Rydberg Constant.

One of the features of Bohr's atom is that the gap between adjacent energy levels gets less as the level number n increases. For example, the gap between Level 1 and Level 2 is 75% of a Rydberg. But the gap between Level 2 and Level 3 is less than 14% of a Rydberg, and between Level 3 and Level 4 is less than 5% of R. And, as we have seen, between Levels 4 and 5 the gap is just 2¼% of R. This makes the energy levels of the Hydrogen atom look like the branches of a Christmas tree – the higher up the tree you go the closer the branches are together, so it is just a small hop for a robin to get from the higher branches to ones just below. But if the robin at the top of the tree sees a worm on the ground at the bottom, it is a big jump to get down to it all in one go; maybe it is safer to hop down a branch at a time (Figure 2.1).

18   The Chemical Cosmos

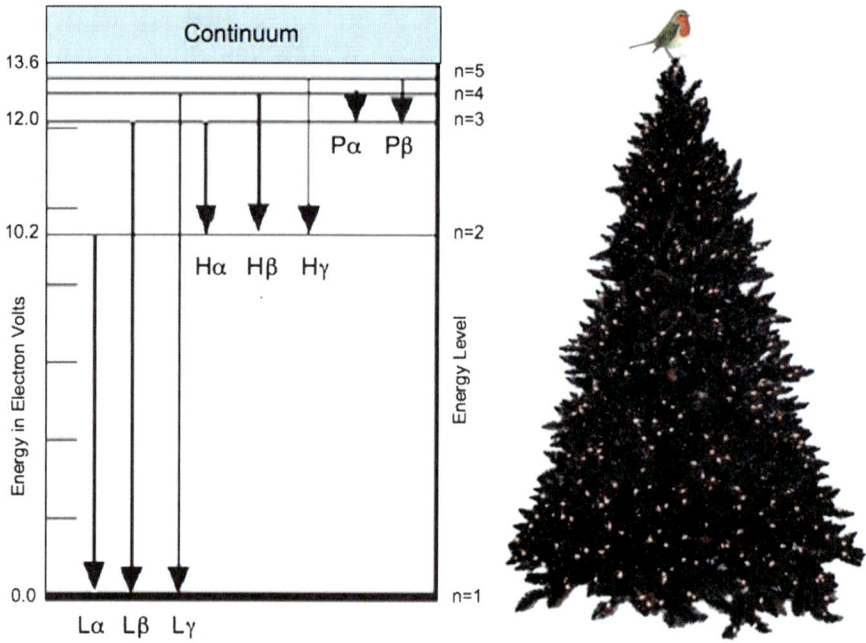

FIGURE 2.1 The energy levels of the Hydrogen Atom: a robin on an Xmas Tree can jump all the way to the lowest branch or hop down branch-by-branch, giving up much less energy per hop: *credit – Steve Miller.*

Back in our early universe, the simplest way to get rid of excess energy is for the combining free electron and atomic nucleus simply to hand it over to just one independent photon. For a Hydrogen atom forming from a free electron and a free proton, and ending up with the electron in the first – and lowest – energy level, that would mean producing an energetic photon with an energy equivalent to R. What goes down can also go up again, however; absolute dictators, for example, know that it is unwise to name a successor who will inherit all that personal power because it is a sure-fire way of getting yourself assassinated. Much better to groom a stable of acolytes each of whom can inherit only some of your powers, and to make sure that they never quite get it together enough to make it worth their while to kill you off. Similarly a Hydrogen atom that had settled down to a comfortable existence with its electron in Level Number 1 might suddenly find its peace shattered by bumping into a photon with an energy equal to R, emitted by a neighboring atom, and ending up re-ionized back into a free proton and a free electron. And if it and all other Hydrogen atoms did likewise and gave up all their ionization energy in one go, stable atoms would not form, and there would be no *Chemical Cosmos* – we would be back to square one.

Although Bohr's structure for the Hydrogen atom is now considered primitive and has been superseded by more detailed modern Quantum Mechanical models, it does, however, serve to demonstrate that recombination does not have to be an all or nothing process. Instead, the recombining atom can proceed from its free-nucleus, free-electron state down to its lowest – Level 1 – energy level by two or more stages, giving off two or more photons each of which has energy less than the Rydberg Constant. So each of these photons is unable to re-ionize its neighbors on its own. Even the largest and final jump – from Level 2 to Level 1 – only has an energy equivalent to ¾ R, a quarter of a Rydberg too little to re-ionize another atom. By the end of the Recombination Era, the universe was sufficiently spread out that the chances of several photons all ganging up on one poor Hydrogen atom to re-ionize it were very few and far between. Neutral atoms could form safely!

Immediately after the Big Bang, the universe was hot and energetic, but very uniform. Even at the end of the Recombination Era, the universe was so "smooth" that only small differences of about one thousandth of a percent show up in the Cosmic Microwave Background Radiation. Nonetheless, by the time that the universe was about 100 million years old – 1,000 times older than it was at the start of the Recombination Era, but very young by comparison with its current 13½ billion years – gas clouds vast enough and dense enough to form 100,000 or even one million stars the size of our Sun were quite common. These enormous gas clouds had been "seeded" by halos of dark matter, cold material composed of exotic particles that interact so weakly with "normal" matter – the kind that we, our planet Earth and our Sun are made of – that they have never been detected. These vast, dense gas clouds are known as proto-galaxies. Although they are very large, these very first galaxies are small in comparison with our own galaxy; the Milky Way is more than a million times more massive than the earliest proto-galaxies. And unlike the Milky Way, or other galaxies that we can see today, such as those in the constellation of Andromeda, proto-galaxies did not yet have stars.

So the task was to form the very first stars. Stars would be the next step towards the rich Chemical Cosmos that we enjoy today.

Stars are themselves huge balls of gas. But even a fairly middling star like our Sun has a density greater than that of the water we drink, and more than a thousand times more dense than the air that we breathe. The gas clouds that formed in the early universe, however, were a hundred billion billion times *less* dense than air. To form the first stars, therefore, meant forming dense clumps within the individual clouds, clumps that would eventually become a trillion trillion times denser than the original cloud. Unfortunately, gas heats up as it condenses, and hot gas tends to expand rather than contract. To have clumps dense enough to make stars meant cooling the gas down sufficiently and rapidly so that gravity had enough time to pull everything together before it became hot enough to fly apart. That, in turn, meant the gas temperature had to get down to just 1,000K or 2,000K. Something had to cool it.

Atoms can cool by radiating photons as their electrons jump down from higher energy levels to lower ones. As hot atoms fly about in the gas with great energy, they can crash into one another. The outcome can be that one atom has its electrons changed so that they hop up to a higher energy level, while its colliding partner looses steam and cools down. The now very excited atom can then cool back down by firing a photon off into deep cold space, where its energy can no longer heat the gas cloud; cooling has been effected. For atomic cooling to work, the colliding atoms have to have enough energy so that at least one of them can have its electron excited. Hydrogen atoms need energy equivalent to a temperature of about 12,000K to push an electron from Level 1 to Level 2, and it turns out that the atoms formed in the early universe are only good at cooling things down for temperatures above 8,000K. That still leaves the gas needing to drop another 6,000K in temperature. Something else is needed – the chemical combinations of atoms known as molecules.

Molecules mean that you really do have a Chemical Cosmos. Starting simple, molecules can grow into more complex creatures. Eventually, they can grow as complicated as the DNA that holds the genetic code for life, so for chemists, the Recombination Era marks the start of the good times. Back then, however, with just a few types of atoms – Hydrogen, Helium, Lithium and Deuterium – there is only so much chemistry you can do, especially when 90% of those atoms are Hydrogen and almost all of the rest are Helium. And making molecules is easier said than done; close to its source, the river of cosmic chemistry is a rather narrow, rocky stream.

By the middle of the Recombination Era, some 200,000 years after the Big Bang, most of the Helium had recombined to form neutral atoms. Much of the Hydrogen, on the other hand, was still in the form of positive nuclei (protons) without their neutralizing electron. Therefore one of the first molecules to form up was a molecular ion, not our chemical guide H_3^+, but a Hydrogen-Helium combination of just one Helium atom and a Hydrogen nucleus, Helium-Hydrogen-plus, denoted HeH^+. This reactive, little molecule then tagged onto whatever Hydrogen atoms had been able to "go neutral", eventually forming our diatomic Hydrogen molecule, H_2, the most fundamental of all molecules.

Unfortunately for these first cosmic attempts at making molecules, the background temperature of the universe was still hot enough to cause them to break up again. Almost as soon as molecules formed, they shook themselves apart, disintegrating too soon for them to play any real part in cooling down the gas clouds to the point where they could start to form stars. To create significant amounts of diatomic Hydrogen molecules meant waiting until more or less the end of the Recombination Era, when both positive and negative Hydrogen ions, H^+ and H^-, could combine directly with neutral Hydrogen atoms to start the formation of H_2, rather than using Helium as a matchmaker.

Even so, astronomical amounts of time were required to convert the gas in the universe from atoms to molecules; making just one Hydrogen molecule per cubic centimeter took over a week. As a result, some 100 million years after the Big Bang, Hydrogen molecules still only made up one part in 400,000 of the Chemical Cosmos. A network of over 40 chemical reactions, involving Hydrogen, Deuterium, Helium, and Lithium, and a variety of positive and negative ions, still only managed to produce ten other molecules at vanishingly small concentrations. Nonetheless, it is at this very early stage of the Chemical Cosmos that our guide,

H_3^+, made its first appearance. At a concentration of just one part in a billion billion, it came in right in the middle of the batting order at Number 6 out of the 11 molecules to be found.

They may be vanishingly small, but even at these concentrations molecules have a significance way beyond their numerical abundance. Once molecules form up, the Chemical Cosmos begins, and the universe can make a start to become the universe we have today, with stars and the possibility of planets, with clusters of stars, and the vast clusters of star clusters that we call galaxies, and even clusters of galaxies! Molecules can accomplish what atoms fail to do.

Like atoms, molecules may also give off photons as a result of changes in the motion of the electrons that surround their constituent atomic nuclei, and also like atoms they can use these changes in electronic motion to cool down to a few thousand degrees. Unlike atoms, however, molecules are made up of several atomic nuclei held together by chemical bonds that can be thought of as little springs, vibrating as a result. As they move through space, molecules can also tumble about like nanoscale circus performers; molecules have spectra that are caused by the motions of the atomic nuclei from which they are made, vibrations and rotations. These motions are, like the electronic motions, quantized; the vibrational and rotational energy of a molecule can only have a certain set of values.

The trick that molecules have is that the energy levels and jumps associated with vibrations and rotations are much less formidable than those associated with electronic motions. If our robin had to jump from branch to branch on the atomic Christmas tree, it only has to hop from twig to twig on the molecular pine. So that means that molecules can still be excited even when the surrounding gas only has a temperature of a few thousand, a few hundred, and even just a few tens of degrees above absolute zero. An atom or molecule hitting another molecule with too little energy to make the electrons jump can still cause the molecule as a whole to change its vibrational or rotational states. Relaxing once more, these hot molecules can then radiate a photon out into space. The vibration-rotation lines of molecules show up all the way from the visible part of the spectrum, at temperatures equivalent to a few thousand degrees, all the way to the microwave, at temperatures of only a few degrees. That means that they can cool at temperatures well below the 8,000K cut-off for atoms.

The question for the early universe was: are there enough molecules? As we have seen, at an age of 100 million years, the universe still only had one Hydrogen molecule, H_2, for every 400,000 Hydrogen atoms. Moreover, on its own, the Hydrogen molecule is actually not a very good radiator, taking something over 10 days to emit a single photon once it has been excited in the first place. So collapsing a cloud of gas to make a star in the early universe was a slow process; with a density of just ten million Hydrogen atoms in each cubic meter of gas (the air we breathe has more than a billion billion times as many) cloud collapse took 15 million years. For nearly the whole of these 15 million years, the gas temperature slowly cooled from about 1,000K to less than 200K, photon by painful photon. Its density increased by a factor of 1,000; now there were ten billion Hydrogen atoms for every cubic meter of gas. Then things started to warm up again; the gas density was high enough that hot atoms could pass on their energy to cooler atoms or molecules before the Hydrogen molecules could radiate it away in the form of photons. The last few thousand years, while the clump got dense enough to form a star, were a constant battle between gravity pulling the gas cloud together and the heating trying to push things apart again.

And what a battle it was.

In the last 10 years during which the gas cloud underwent its final gravitational collapse to form a star, the density had risen to ten billion billion molecules per cubic meter – still over a million times fewer than in the air that we breathe, but a trillion times denser than when the cloud started to collapse all those 15 million years previously. Fighting this collapse, the temperature had now increased again to over 1,000K. Hydrogen molecules worked hard to keep the gas cool, but during this final 10 years only one in every two Hydrogens managed to emit a single solitary photon. With six billion billion Hydrogen molecules in every cubic meter of gas, shortly before the first stars "turned on" the cooling rate was a miniscule one billionth of a watt. Tens of cubic kilometers of gas were only giving out as much energy as a single household light bulb!

Yet it was enough and stars did form. Once formed they could enrich the Chemical Cosmos as never before.

So if Hydrogen molecules are not very good coolers, one might ask if there was anything else. The answer is not really. The fact that Hydrogen molecules were so much more abundant than any other molecule meant that they dominated the cooling of the gas cloud. They were not entirely alone, though; a Hydrogen atom can team up with an atom of heavy Hydrogen, called Deuterium, to form a molecule denoted HD instead of our normal Hydrogen molecule, H_2. HD is a much better cooler per molecule than H_2, and if the gas became shocked and compressed faster than the speed of sound, it could contribute considerably to the subsequent cooling. There was also a critical period of some 10,000 years while the density of our collapsing gas cloud increased from around 10 billion to more than 100 billion atoms per cubic meter; the temperature was rising sharply, and our cosmic guide, H_3^+, showed what it could do. Although there was only one H_3^+ molecule for every billion of H_2, it managed to contribute more than 1% of the total cooling. With that effort, each H_3^+ molecule showed it is at least ten million times more effective at cooling as its neutral parent, a property that will be important later on in our guided tour.

For now, however, along with our guide, we are on our way downriver to the stars.

3. Shooting the Rapids: The Life and Death of the Earliest Stars

At 7:35 Universal Time on February 23, 1987, they started to arrive. Too faint to be detected for another 3 hours, they were the first messengers announcing that something momentous had happened in a galaxy not so far, far away. A giant blue star, originally some 20 times more massive than our Sun, had finally found that its own gravity was too much to bear. It had collapsed in on itself and then rebounded in an enormous supernova explosion. Traveling at the speed of light, photons from this massive explosion had reached Planet Earth some 150,000 years later. Over the course of the next day, Supernova 1987A, as it became known as, had brightened by more than 500 times until at last it was spotted by Ian Shelton at the Las Campanas Observatory in Chile. Amateur observers helped to fix the exact time at which the explosion happened, and their results were later confirmed by the detection of neutrinos (will-o-the-wisp particles that interact with matter very rarely and weigh so little that they can travel at almost the speed of light) that were produced in the dying moments of the star that produced the supernova.

Supernovae are the drama queens of the cosmos. While other stars live quiet, unassuming lives and end their days as dense glowing embers, supernovae literally go out with a bang, pouring their hearts out into the surrounding universe as they go. We should be grateful to them, however, because they give us the richness of the Chemical Cosmos without which our chemically complicated lives would not be possible.

The alchemy of the European Middle Ages, although infinitely more rich and subtle than it is often given credit for, was in one way or another based on a few elements or principles. These were often cast as the four Aristotelian elements: Earth, Water, Air, and Fire, or more prosaically, salt, sulfur and mercury.

During the scientific revolution of the seventeenth century, much of the work of the alchemists was shown to be of a simplistic and quasi-magical nature. Robert Boyle, one of the founders of the Royal Society, England's premier scientific academy, produced a blow-by-blow demolition of these simplistic ideas, published in 1661 as *The Skeptical Chymist*. While not breaking from the principles of alchemy, a characteristic he shared with Sir Isaac Newton, Boyle came to the conclusion that if there were just a few fundamental elements, they were mechanical, not chemical: matter, motion and rest. For the rest, he demonstrated that almost all of the elemental variations on an Aristotelian theme were compound materials that could be further broken down, altered and recombined.

During the course of the eighteenth and nineteenth centuries, chemists built up an impressive catalogue of chemical elements and noticed some important regularities about them. The two lightest elements were Hydrogen, H, and Helium, He. Hydrogen was a reactive gas that exploded when mixed with Oxygen; Helium was a gas, but it was almost totally unreactive and inert. Next in the list of elements by weight came a series of 8: Lithium, Beryllium, Boron, Carbon, Nitrogen, Oxygen, Fluorine, and Neon. Then another octet: Sodium, Magnesium, Aluminum, Silicon, Phosphorus, Sulfur, Chlorine, and Argon. The first 18 elements all followed each other in their atomic weights, but not in ways that were always easy to understand. If Hydrogen was assigned a weight of 1, for instance, why was Helium 4, Lithium 7, Carbon 12, or Oxygen 16? And, even more perplexingly, why was Chlorine 35½?

Patterns nonetheless emerged. The first elements in the octets were reactive metals, Lithium and Sodium, which burned fiercely on contact with water and tended to combine with Hydrogen on the basis of one unit of metal with one unit of Hydrogen. Next in both octets came two more metals, Beryllium and Boron in the first and Magnesium and Aluminum in the second. All four were reactive metals, but Beryllium and Magnesium combined with two units of Hydrogen, while Boron and Aluminum combined with 3. In the middle of the octets, Carbon and Silicon were not really metallic, but both combined with four units of Hydrogen.

In the first octet, Nitrogen and Oxygen were both gases, while in the second octet Phosphorus and Sulfur were both non-metallic solids. Yet both Nitrogen and Phosphorus combined with three units of Hydrogen, indicating they were from the same chemical family, while Oxygen and Sulfur went for two Hydrogen units – again, a family resemblance. At the end of the octet, Fluorine and Chlorine were both highly corrosive gases, which formed strong acids with one unit of Hydrogen only, while Neon and Argon, like Helium, were almost totally inert.

So we have a nice set of family resemblances, and by putting these and other elements into a periodic table, the Russian chemist Dmitri Mendeleev was able to predict that new elements might be discovered and what the properties of their chemical compounds might be. In particular, Mendeleev found that element 21, which should have been next to Calcium, was missing; therefore he put in a new element called EkaBoron and predicted its properties. Similarly elements 31 and 32 did not seem to be around in the current inventory, leaving gaps in the Boron/Aluminum and Carbon/Silicon families, respectively. Mendeleev's predictions were realized in the shape of Scandium for element 21, Gallium in position 31 and Germanium in position 32. The patterns made learning chemistry much easier and brought a certain degree of satisfaction with them. For some chemists, they gave rise almost naturally to the idea that the chemical elements came in the form of atoms, and that the ratios in which atoms of different elements combined was due to the fact that individual atoms were joining together according to their different powers to combine.

J.J. Thomson's discovery of the electron, which we shall discuss later on, and its subsequent incorporation into the Rutherford-Bohr model of the atom and later models that made use of modern Quantum Mechanics, demonstrated that the combining powers of different atoms with their own kind or those of other elements were due to changes in the electronic structure and properties of those atoms as they combined. More particularly, the way in which atoms joined hands in these bonds involved sharing electrons in such a way that each atom could complete an octet of electrons – at least that is how it worked for elements heavier than Hydrogen and Helium.

Sodium, the first element of the second octet, donated an electron to Chlorine, second to last element of the second octet. Sodium, Na, became the cation Na^+, with the full octet of electrons belonging to the element before it, Neon, last member of the first octet. Chlorine, Cl, became the anion Cl⁻, with the full octet of the last member of the second octet, Argon, and together, Na⁺ and Cl⁻ happily formed common salt. Carbon, C, needing four electrons for an octet, shared two electrons with two Oxygen, O, atoms, each of which needed two more electrons to reach an octet. By sharing back and forth Carbon and both Oxygens reached their octets and merrily went on their way together as CO_2, Carbon Dioxide.

So much for the combining power of atoms, but how now to account for the weights of the elements? In some ways the early work on radioactivity by Rutherford and others made things worse. True, Rutherford's atom had its weight concentrated in a heavy nucleus, surrounded by electrons that, individually, weighed less than 1 part in 2,000 of a Hydrogen atom (1 part in 1,836, to be more accurate). Hydrogen, with just one positively charged proton in the nucleus, had a weight of 1. So why was the Helium nucleus – the alpha particle that sometimes cannoned back off a thin gold foil – of weight 4, when it was clear that it had only two positive charges in its nucleus?

Isotopes have a bad image, almost always they are associated with the word *radioactive*, and they conjure up images of atomic mushroom clouds, failed power stations and terrorist plots. Yes, the isotope of Uranium known as U-235 is the stuff of atomic bombs, and much of north-west Europe was affected by radioactive Cesium isotopes raining out onto agriculturally productive land as a result of the 1986 Chernobyl power station explosion. Yet isotopes are also used for peaceful and benign purposes ranging from smoke detectors and dating archaeological remains, to the treatment of cancer.

At the turn of the nineteenth century, it seemed as if scientists were discovering strange new phenomena almost daily. In 1887, the German physicist Heinrich Hertz had discovered that intense beams of light falling on a metal surface would cause electrons to be ejected – an effect that was later characterized as the photo-electric effect by Hertz's Hungarian assistant, Philipp Lenard. In 1905, in what was his *annus mirabilis*, Albert Einstein had shown that this was proof that light consisted of particles, called photons, with an energy that depended on the color, or frequency, of that light. The electrons in the metal surface were knocked out as a result of being hit by light particles, but the nuclei of the metal atoms were too heavy to be affected. So it was a great surprise when in the 1920s Irene Joliot-Curie and her husband, Frederic, found that radioactive rays from the element Polonium, discovered by her famous mother Marie Curie, could knock Hydrogen nuclei (protons) out of paraffin wax. The Polonium rays the Joliot-Curies were using were electrically neutral, so they could not be electrons or protons, leaving light particles, or photons, which were known *not* to be able to kick protons out of solids. It fell to James Chadwick, working with Rutherford at Cambridge, to show that the rays coming out of Polonium were in fact a new particle altogether.

This particle was the neutron. It had approximately the same mass as a proton – 1,839 times the mass of the electron rather than 1,836, but so similar as hardly to matter when it came to working out atomic weights. Unlike the proton, however, the neutron was not electrically charged; it was neutral, and now the composition of atoms could be understood. For elements heavier than Hydrogen, Rutherford's nucleus consisted of both protons and neutrons. Surrounding the nucleus was a cloud of electrons, and – in a neutral atom – there were exactly the same number of electrons in the cloud as there were protons in the nucleus, but the neutrons were not involved in maintaining electrical neutrality as they did not have an electric charge.

Hydrogen, the first element, had a very simple nucleus consisting of just one proton, so the Hydrogen atom consisted of one proton surrounded by an electron "cloud" of just one electron. Helium, element number 2, had a nucleus consisting of two protons and two neutrons, giving it an atomic mass of four, surrounded by a cloud of two electrons. Element 11, Sodium, for example, had a nucleus consisting of 11 protons and 12 neutrons, giving it a mass of 23, surrounded by a cloud of 11 electrons to maintain neutrality. But that still left the question of how was it that Chlorine had a mass of 35½? After all, protons and neutrons both had a mass on the atomic scale of 1, and you could not get half a proton or half a neutron.

It turns out that normal Chlorine gas consists of a mixture of two main types of Chlorine atoms. One type consists of atoms that have 17 protons in the nucleus and 18 neutrons, surrounded by a cloud of 17 electrons. This atom of Chlorine has a mass of 35. The other main type has, again, 17 protons in the nucleus and 17 electrons in the cloud, but 20 neutrons are packed in with the protons in the nucleus rendering the overall weight of this type of Chlorine atom 37. And because the lighter mass-35 Chlorine is three times more abundant than the heavier mass-37 Chlorine, the *average* atomic weight of normal Chlorine gas is 35.5, or 35½. By contrast, Deuterium, the heavy form of Hydrogen which has one proton and one neutron in its nucleus, only occurs at one D for every 50,000 normal Hydrogen atoms, making the average atomic weight of Hydrogen very close to 1.

From this we can see that the nature of an atom, or what species it is, is determined by the number of protons in the nucleus: 1 for Hydrogen, 2 for Helium ... 17 for Chlorine, and so on, but the *isotope* depends on how many neutrons accompany the protons in the nucleus. In the case of Chlorine, there are known isotopes with masses ranging from 28 to 51, corresponding to anything between 11 and 34 neutrons accompanying the 17 protons in the nucleus. Yet the chemical atom is the same – Chlorine, element 17. Often the type of atom and which isotope are represented in the fashion $_Z^A Y$, where the subscript Z before the chemical symbol Y indicates the element number (the number of protons in the nucleus), while the superscript A before the chemical symbol Y indicates the total atomic mass, given by the number of protons plus the number of neutrons in the nucleus. So normal Hydrogen is $_1^1 H$, the most common isotope of Helium is $_2^4 He$, and the two most common isotopes of Chlorine are represented as $_{17}^{35} Cl$ and $_{17}^{37} Cl$.

Isotopes are simply atoms of the same chemical, with the same element number, Z, and the same number of protons in their nucleus, but with different atomic masses, A, as a result of having different numbers of neutrons in the nucleus. Carbon has three common isotopes, and the most common is Carbon-12, with six protons and six neutrons in the nucleus, making up nearly 99% of all Carbon atoms. This is a very stable isotope, as is the next heaviest Carbon-13, which has six protons and seven neutrons, but Carbon-14, with six protons and eight neutrons, is radioactive and unstable. In just under 6,000 years, half of all the Carbon-14 in any organic object will have decayed away, which makes it ideal for dating archaeological objects.

Oxygen has three common isotopes. Oxygen-16, with eight protons and eight neutrons, makes up 997 atoms out of every 1,000, with the heavier Oxygen-17 and Oxygen-18 making up (almost all of) the rest. Among the heavier elements, multitudes of isotopes are common, so what about the lightest elements? It turns out that element 2, Helium, which normally occurs with a mass of four, showing that it has two protons and two neutrons in the nucleus, can also exist as a lighter form of Helium-3, which has only one neutron to accompany the two protons in the nucleus. Helium-3 atoms only make up one atom in a million of the overall Helium on Earth but because they are radioactive, there are now serious discussions about using Helium-3 as a way of meeting our energy needs without contributing to global climate change.

By the time that George Gamow was proposing his 1948 Big Bang theory, more than 100 chemical elements, many with several isotopes, were known. Gamow's Big Bang envisaged fundamental nuclear particles – protons and neutrons – crashing together with such energy that they stuck to form atomic nuclei heavier than Hydrogen. With the temperature still around three billion degrees, and the time just about 3 min after the Big Bang, the nuclear reactions began. The first to happen was the collision of a proton and a neutron to form Deuterium, heavy Hydrogen, and denoted $_1^2D$. Not much of that Deuterium remained, however, partly because its formation was quickly followed by further collisions that formed either another type of heavy Hydrogen nucleus, that of Tritium $_1^3T$, or the light form of Helium $_2^3He$ – Helium-3. A further collision, with a proton in the case of Tritium, or a neutron in the case of Helium-3, gave rise to normal Helium, $_2^4He$.

Next in line was element 3, Lithium, denoted $_3^7Li$, that could be formed in small amounts from collisions between the existing particles and atomic nuclei. Yet Gamow's Big Bang stalled at this point. Crucially, element 4, Beryllium, refused to form a sufficiently stable nucleus, although it should have been relatively easy; two normal Helium nuclei $_2^4He$ had the required four protons and four neutrons to make the Beryllium nucleus, denoted $_4^8Be$, or Helium could combine with its light cousin to form $_4^7Be$. They just would not hang together, though, and before the bottleneck could be overcome, the universe, still not much more than half an hour old, cooled to a temperature too low for nuclear synthesis of the elements to continue. So much for the Big Bang as a way of making heavy elements! Another way, therefore, had to be found.

Stars, massive spheres of hot gas, live in constant danger of collapsing in on themselves. Indeed, they are born when vast interstellar gas reservoirs start to condense around clumps of the clouds just that bit denser than their surroundings. As the gas cloud collapses, the center becomes extremely hot, with a temperature of ten million degrees or more. That is enough to get Hydrogen to burn, as in the Big Bang, creating Deuterium and Helium and generating enough heat to make the gas expand at the same rate that gravity is causing it to collapse. A dynamic stand-off is reached, and for our Sun, this stand-off has already lasted about 4½ billion years and will last about the same time again. All the while, a nuclear furnace is burning at the center of the Sun, so scientists began to turn to stars as agents to synthesize the heavier chemical elements after the Big Bang had cooled. Stars still had to get past the bottleneck at element 4, Beryllium, if they were to produce the elements of life; the rapids of element formation still had to be negotiated.

There is a philosophical principle known as the Anthropic Principle that is often used to explain why the universe is as it is. In its *strong* form, it proposes that various physical and chemical conditions in the universe have been set just the way they are so that we humans, intelligent, self-conscious, reflexive beings, can exist. More often than not, some form of deity – albeit a "super-mathematician" – is invoked to explain who does the setting (and why). In its *weak* form, however, the Anthropic Principle can be used to deduce things about the universe from the fact that we do exist. In this instance, the fact that we are Carbon-based life-forms means that Carbon just has to exist, and that somehow it has to be possible to get past the bottleneck at element 4 on to elements 5 (Boron), 6 (Carbon), 7 (Nitrogen), 8 (Oxygen) and so on.

In 1954, English Astrophysicist Fred Hoyle made the bold proposal that chemical synthesis in stars did not wait for significant amounts of element 4, Beryllium $_4^8Be$, to build up from two colliding Helium nuclei, before it teamed up with another Helium $_2^4He$ to make Carbon $_6^{12}C$. Instead, Hoyle proposed, three Helium nuclei could come together more or less simultaneously to form Carbon all at once. For this to happen, however, something had to be done about the excess energy that three colliding Helium nuclei would bring to the party. If not, the three would exchange a few pleasantries and then go off on their separate ways again, and no Carbon would be formed. Hoyle's solution was to suggest that there would be a special energetic state of the Carbon nucleus that would absorb this energy. He persuaded atomic physicist Willie Fowler of the Carnegie Institution in Washington to search for this special state of Carbon: Fowler found it and the bottleneck was overcome. Carbon could, after all, be made by nuclear synthesis in the center of stars. Carbon *could* therefore exist.

Hoyle and Fowler called the three-Helium collision method of making Carbon the "triple alpha" process, since three alpha particles came together to produce it. Together with colleagues Margaret and Geoffrey Burbidge of the California Institute of Technology, they then worked out how the rest of chemical elements could be made. Their paper *Synthesis of the Elements in Stars*, by E. Margaret Burbidge, G.R. Burbidge, William A. Fowler and F. Hoyle, was published in the *Reviews of Modern Physics* in 1957, and is one of the classics of post-war astrophysics.

The stars in the night sky may look unchanging. For example, the bright constellation of Orion the Hunter is present winter after winter, and the Summer Triangle of Vega, Deneb and Altair keeps its shape year in and year out. That is on a timescale measured by individual humans, though, and over periods of thousands, millions and billions of years, the stars are forever changing. They evolve from youth to old age, forming out of dense gas clouds and going through very active infancies, known as a T-Tauri phase after the young stars in the constellation of Taurus the Bull. They settle down onto what is known as the Main Sequence to potter along for most of their days; our Sun is currently on this Main Sequence. It has been for some 4½ billion years already and will stay on the Main Sequence for about as long again.

During its time on the main sequence, the Sun is burning Hydrogen to form Helium, via Deuterium, making use of individual protons and neutrons, and finally Helium-3 collisions to form the stable $_2^4$He normal isotope. Present day stars that are greater than 50% more massive than our Sun, and which already have significant amounts of Carbon in their centers, can make use of another route to form Helium from Hydrogen. This involves normal Carbon-12, $_6^{12}C$ absorbing a proton, $_1^1H$, to form a radioactive isotope of Nitrogen, $_7^{13}N$. Through a series of further nuclear reactions, the atomic nucleus builds all the way to Oxygen, $_8^{16}O$, before breaking down again. This eventually produces $_2^4He$ and returns the $_6^{12}C$ that initiated the chain, to start all over again. During this process, four protons have been used up, two providing the two positive charges needed in the Helium nucleus, while the other two have turned into neutrons, each emitting particles of antimatter, positrons, or positively charged electrons, in the process.

Towards the end of their lives, stars bloat into giants, like the Red Giant Aldebaran in Taurus or Arcturus, the Hoku Lea or Happy Star of the Polynesian Pacific voyagers who founded Hawaii. They may also expand into supergiants, like Betelgeuse in Orion or Antares (the rival of Mars) in the Scorpion, Maui's fish hook. During the final phases of their lives, these giant stars produce Carbon by burning Helium in the Burbidge, Burbidge, Fowler and Hoyle triple alpha process. These stars can also build atoms of Oxygen and Nitrogen, and other heavier elements, making use of Helium nuclei and neutrons to do so. Later, they also make heavier elements by rapidly burning Carbon and Oxygen. For example, two colliding nuclei of Carbon-12 can make either element 10, Neon, or element 11, Sodium, in one go, giving out a Helium or a Hydrogen nucleus, respectively. In their final throes, giant stars hurl their outer layers into space, vast amounts of gas enriched with the heavy elements of life, to form what are confusingly known as planetary nebulae. This rich elemental soup goes on to form the next generation of clouds from which new stars will eventually form, while the core of the old star contracts into a dying ember like a White Dwarf, or even a neutron star, whose overall density is that of an atomic nucleus.

To become first a Red Giant and then a White Dwarf is the fate that awaits our own modest Sun, but large stars with masses nine times or more than that of our Sun live short, hot lives measured in millions, rather than billions, of years. They burn their Hydrogen fuel at a much faster rate, pouring their energy out into space as bright blue light. Supernova explosions happen when such a star has used up all of the nuclear fuel at its center. Not only has it burned up its Hydrogen, it has also eaten up its stocks of Helium, Carbon, Neon, and Oxygen in rapid succession. That only leaves

Silicon and Sulfur, which keep the star alive for just a couple of more days. Once heavy elements like Iron get to dominate the star's center, nuclear burning, adding more protons and neutrons, requires energy rather than releasing it, and there is nothing left to fight off the effects of gravity.

Collapse takes over, and the temperature at the center rises to ten billion degrees, a thousand times the original temperature. Some of the Iron is ripped apart once more as the photons of light desperately seek a way out of the morass, while the density becomes so great that negatively charged electrons are actually captured by the positive protons to form the beginnings of a neutron star. The blast that accompanies all of this fires a mass of heavy elements – from Carbon all the way through to the radioactive elements like Uranium – out into space. This explosion is enormously large and is equivalent to over 30 billion billion billion megatons of high explosive. Planets, such as those in our Solar System, caught in the blast from such a supernova would be totally obliterated.

In all probability, the very first stars to form in the universe (the stars that were to populate the original proto-galaxies) were either very large or very small – just after star formation began there would have been a mixture of each. Depending on whether the cloud could be kept cool only by molecular Hydrogen, H_2, or if it could be cooled by other molecules, such as the Hydrogen-Deuterium HD version or our guide H_3^+, it would either break into clumps that were more than a hundred times more massive than our Sun or clumps that were only a small fraction of the Sun's mass. In the first case, the star that formed would be a supergiant anywhere between several tens and a thousand times more massive than the Sun. This supergiant's short life would be characterized by an outpouring of energy, ultraviolet radiation so intense that the rest of the cloud would be broken back into individual Hydrogen and Helium atoms, and those, in turn would have been re-ionized to atomic nuclei and electrons. No further star could form while the marauding supergiant held sway, but that would not be for long. Such stars live just a few million years before "going supernova," and then pretty much everything they have and have made gets thrown back out into space. That included the very first atoms heavier than Lithium: the first Carbon, the first Oxygen, the first Nitrogen, the first elements required for life.

36   The Chemical Cosmos

FIGURE 3.1 Twenty years after the explosion was noticed on Earth, the Hubble Space Telescope took this image of Supernova 1987 A. The rapidly expanding gas cloud from the explosion has crashed into gas and dust surrounding the original star, creating a bright "ring of pearls" about a light year across. Fainter red rings, lit up by the explosion, can be seen at the top and bottom of the picture: *credit NASA/ESA/Hubble Space Telescope Science Institute.*

Although the very earliest supergiants, the stars that seeded the universe with all the elements heavier than Lithium, are long, long gone, studying supernovae still enables us to get a handle on those early days. So following a supernova explosion from the word "go" was high on the wish-list of astronomers. Supernova 1987A was ideal: it was in the Large Magellanic Cloud, or LMC, a small galaxy around 1½ million billion kilometers from the Sun, very close in astronomical terms but far enough away not to pose a threat to Earth (Figure 3.1).

The International Astronomical Union (IAU), which unites astronomers worldwide, issues rapid circulars to alert its members to important events. Supernova 1987A most definitely qualified as an important event and Ian Shelton's IAU Circular Number 4316 galvanized observatories worldwide to study it. The resulting "light curves" showed how Supernova 1987A brightened in the night sky, before fading away again over the hours, days, weeks, and years after Shelton's announcement. UV radiation first showed a temperature of some 14,000K in vast gas cloud that the supernova generated, cooling rapidly as it expanded at nearly 20,000 km/s, 6% of the speed of light. As time went on, beautiful images showing concentric rings of pink gas – shock-heated Hydrogen atoms – of enormous proportions became a staple of talks, newspaper and magazine articles, and TV broadcasts, illustrating the widespread fascination with this cosmic explosion.

The spectrum of Supernova 1987A was measured in every wavelength possible from shortward of the ultraviolet to the infrared and beyond, so that the chemical makeup of the gas could be analyzed. How much of what was produced and when it was produced was crucial to understanding the details of the supernova process as it evolved in time. It was also important to understand just what kind of a star had given rise to Supernova 1987A, for until this event, conventional wisdom had been that red supergiant stars, like Betelgeuse in Orion, were the ones that were expected to go supernova.

It turned out that Supernova 1987A's progenitor was blue, and not red, and that took some explaining. The most likely reason was that the star that eventually exploded had already thrown about a quarter of its original mass into space to form an enveloping cloud of Hydrogen-rich gas. Billions of kilometers distant from the dying star, this gas was not involved in the original supernova explosion. It would, however, soon be caught in the blast as it traveled out at thousands of kilometers per second, reaching out into space more than 40 billion kilometers in about 100 days. How, then, would the explosive shock of the star react with the large gas cloud it found there?

The motto of the UK's Royal Astronomical Society is *Quicquid Nitet Notandum* – all that shines from the heavens is noteworthy. That does not mean to say that all that shines from the heavens gets noted; well, at least, not immediately. Peter Meikle and his group at Imperial College in London's regal South Kensington district had been measuring and monitoring the infrared spectrum of Supernova 1987A using the Anglo-Australian Observatory in Epping, New South Wales. Their campaign had started in the middle of the supernova's third week on March 13, 1987. It had run a full year, marking the steady brightening until mid-June, after which the afterglow of the supernova explosion started to wane and cool until the dying embers were 50 times fainter than at their peak.

The supernova's light curves provided vital information about the evolution of this stellar explosion. Over the first 3 days, the ultraviolet radiation given out by the supernova declined by a factor of 1,000 as the hot gas thrown out by the explosion cooled rapidly as it expanded into space. By the time the explosion was three months old, however, it was clear that the main source of radiation was due to the gas being reheated by the decay of radioactive Cobalt. Even more vital was the way in which Meikle's group was able to use observations to pick out just what had been produced in the explosion. The earliest measurements showed up Carbon, Magnesium, Calcium, and Strontium at high intensities, as well as several lines of atomic Hydrogen, from which they deduced that the debris from the explosions was expanding at well over 5,000 km every second, nearly 2% of the speed of light. As time continued, Helium, Oxygen, Sodium, Silicon, Sulfur, Potassium, and Calcium could all be seen in the mix. There were even traces of Iron and Cobalt, some of the very last elements to be synthesized by the core of the supernova before it finally blew itself to pieces.

By mid June, other features were beginning to appear in the mess of fingerprints that made up the Supernova 1987A's spectrum. When Meikle's group looked again on September 3, 1987, it was clear: the new features were molecules. Simple molecules to be sure – Carbon Monoxide and Silicon Oxide, Carbon Sulfide – but molecules nonetheless, a clear sign that the expanding gas cloud was finally cooling down enough for atoms to begin joining hands together, without shaking themselves apart. By now, Supernova 1987A's gas cloud had expanded to such an extent that it would have totally engulfed all of the Solar System's major and minor planets, as well as the vast Kuiper Belt from which many of our comets come. Meikle's spectra were reported at conferences and

Shooting the Rapids: The Life and Death of the Earliest Stars

FIGURE 3.2 Alex Dalgarno: *credit – Harvard University, Institute for Theoretical Atomic, Molecular and Optical Physics.*

finally published a year later in the *Monthly Notices of the Royal Astronomical Society*, for all to look at and to ponder over their significance.

Sometimes the shortest distance between two points is not a straight line, at least not where information is concerned. Across the Atlantic Ocean, Harvard's Alex Dalgarno, the legendary former editor of the letters section of the *Astrophysical Journal* who is reputed to have read and understood every one of the articles that crossed his editorial desk, was working with his colleague Stephen Lepp on how Supernova 1987A was making its molecules. Dalgarno was part of a stable of post-war atomic physicists spanning Belfast and London, who had helped to put astrophysics on a sound atomic and molecular footing. As a Brit, Dalgarno made frequent visits to his former homeland to organize and attend conferences and to give seminars (Figure 3.2).

At one of these seminars, at University College London, Dalgarno chose to share what he had been doing on making molecules in the Supernova 1987A gas cloud with others whose interests lay in the chemistry between the stars. He showed Meikle's results and how his chemical models were able to explain the presence of Carbon Monoxide and the other molecules whose fingerprints had been found in the infrared spectra. Good detectives take pains to identify as many as they can of the fingerprints they find at crime scenes, and then to label the rest with "unidentified", or "?". That is what

FIGURE 3.3 Jonathan Tennyson: *credit – University College London, Department of Physics and Astronomy.*

Meikle and then Dalgarno had done, and there were two key "?"s, at the wavelengths of 3.41 and 3.53 microns (μm).

Among Dalgarno's audience that day ware the molecular physicist Jonathan Tennyson, descendant of the Victorian Poet Laureate, Alfred Lord Tennyson, and members of his group (Figure 3.3). Tennyson's group was responsible for calculating the first really accurate spectrum of our guide H_3^+, and had used their calculations to good effect (as we will discuss later) to identify its fingerprints in the giant planet Jupiter. Tennyson himself had not long returned from Hawaii, where he had been using precisely those ?-mark wavelengths to take images of Jupiter's aurorae. Was it a coincidence? One wavelength the same, maybe, but two was a bit much. After the seminar, Dalgarno, Tennyson and excited colleagues sat down to take a closer look at the details of the Meikle spectra. The two key wavelengths were just about discernible in the mid-June spectra, when SN1987A was 110 days old. They were very clear at Day 192, September 3, 1987, and they remained visible all the way through Meikle's first year's data, fading along with the overall brightness of the afterglow, closely following the pattern shown by the Hydrogen lines. It was our chemical guide at work. H_3^+ was putting in its first ever appearance outside of the Solar System, not even in our own Galaxy but in the neighboring Large Magellanic Cloud.

To be seen from the LMC, a distance of 145 million, million kilometers, the amount of H_3^+ produced had to be large, and it had to be hot. The fingerprints in Meikle's spectra were clear, but a bit messy, so Tennyson's group estimated that as much as 100 billion, billion tons of H_3^+ with a temperature between 1,000 and 2,000 K had been made to account for what was seen on September 3, 1987. The gas was also being driven across the LMC at high velocities, at thousands of kilometers per second. Accounting for so much of our chemical guide was a problem, because the gas that came out of Supernova 1987A was, by definition, rich in heavy elements. H_3^+, while stable, is reactive; if the chemical soup is rich in Carbon, Nitrogen and Oxygen compounds, our chemical guide does not stand a chance but meekly hands over its proton and goes back to being boring old diatomic H_2 again, albeit with the satisfaction of knowing it has really got the reaction pot boiling.

It was even difficult to get H_3^+ to form at all, according to some models of the supernova, because the gas was still too hot for diatomic Hydrogen to hang together, even if more stable molecules like Carbon Monoxide did. Stephen Lepp and Alex Dalgarno put their heads together and came up with the only possible explanation – most of the molecules that showed their fingerprints in Meikle spectra were being formed in the hot, rapidly expanding gas that came directly from the supernova, but not H_3^+.

As we have seen, before it finally "went supernova", Supernova 1987A's star had been a hot blue star, many times the mass of our Sun. Towards the end of its lifetime, when it was at least grumpy enough to draw its pension but not yet explosively angry, the star had undergone some rather major surgery, resulting in it shedding a large envelope of Hydrogen-rich gas. Such stars are common, and although not visible to the naked eye, give rise to some of the most stunningly beautiful objects in the sky, known as "planetary nebulae". The Cat's Eye nebula, for instance, known by its catchy New Galactic Catalogue number of NGC6543, shows two sets of overlapping concentric rings of green, red and blue-white gas due to the presence of Oxygen cations and Hydrogen atoms, lit up by the powerful radiation of the central star.

Lepp and Dalgarno argued that the identification of H_3^+ by Tennyson's group could only mean that the explosive, heavy-element-rich gas from the core of Supernova 1987A had caught up with the slower moving, Hydrogen-rich envelope the star had shed much earlier. The resulting shock front had heated and ionized diatomic

Hydrogen molecules to start the reactions to form our chemical guide, but the heavy elements had not mixed into the old gas envelope to react with H_3^+ before it could signal its presence to the outside world. The prestigious journal *Nature* was convinced, and published the Harvard-London work as one of its letters in 1992. Not everyone else was convinced, though, and there were murmurings; articles called the work "tentative", but in the years that followed every argument against their model was swatted away emphatically by Lepp and Dalgarno. No other "criminals" were found to explain Meikle's 3.41 and 3.53 μm fingerprints either.

Back in the early universe, many of the stars that formed were much more massive than the star that exploded to form Supernova 1987A, and their exact fate depended critically upon their initial mass. For those between about 9 and 50 times the mass of the Sun, Supernova 1987A, known as a Type II supernova (Type I supernovae are created by smaller stars), would be an ideal model. Between about 50 and 100 solar masses, stars may fail to explode because gravity works too fast and turns them into Black Holes before this can happen. Their destiny then is to linger as a dark menace, visible perhaps as a result of the special radiation they emit, named for the British astrophysicist, Stephen Hawking.

Above 100 times the mass of the Sun, the star may end its days in what is called a "pair instability" supernova, creating vast numbers of pairs of electrons and their anti-matter partners, positrons. If the mass of the star is less than 130 solar masses, then this process occurs in pulses, throwing gas out into space until the star has lost enough mass to become a "normal" supernova. Above 130 solar masses, the outcome is much more explosive and leaves nothing behind. A supernova spotted in 2006, Supernova 2006gy, could have been a more recent version of such a beast. Its progenitor star was about 150 times as massive as our Sun, and exploded some 238 million years ago, at a time when, on Earth, the Triassic Period was beginning to usher in the age of the dinosaurs. Supernova 2006gy let rip an explosion that was maybe as much as 200 billion billion billion megatons, and tore the parent star to shreds, emptying everything out into space. Finally, the largest class of star, those more massive than 250 times our Sun, collapse once more as Black Holes.

Shooting the Rapids: The Life and Death of the Earliest Stars 43

We know that significant numbers of the stars in the early universe were of the right size to go supernova, and to enrich the cosmos with the chemical elements they had synthesized by nuclear reactions. Indeed, it has proved possible to detect simple, heavy-element-containing molecules such as Carbon Monoxide, CO, dating from a time well before the universe was 1 billion years old, or a mere teenager; Water, for example, has now been detected from a Black Hole formed "just" 2½ billion years after the Big Bang. Given the energetic nature of stars that go supernova, it is clear that in the early universe some of the gas must have been ionized like the cloud surrounding Supernova 1987A and our guide H_3^+ must have been there to witness the earliest of supernova explosions, at least at the start. Whether it survived and shone like its fellow molecules more than 13½ billion years later in SN1987A, we shall probably never know for sure, for the afterglow would now be infinitely fainter than the Cosmic Microwave Background Radiation, and well beyond the reach of at least this generation of telescopes and their instruments.

What, then, of the early stars that did not go "supernova", formed from the smaller clumps of gas that could also have formed in the early proto-galaxies? The very earliest of these – like their supergiant siblings – would have been composed of just Hydrogen and Helium. Like small stars today, they would have had a long infancy and an even longer adulthood, an adulthood so long that some of them may even be around today, in or close to our own galaxy. These Population III stars (astronomy names star generations backwards, the youngest being Population I, intermediate generation stars are Population II, and the oldest are Population III) being small have outer layers cool enough for molecules to form up and remain stable. Those molecules would be the usual suspects – H_2, HD and Helium-Hydrogen mixes, but there would also be H_3^+, and enough of it to affect the way in which the star lived its life.

At the core of a star generating energy by nuclear burning, high energy photons are being rapidly produced. These are gamma rays, with wavelengths typically of a few percent of a nanometer. But stars do not emit all of their energy as gamma rays, or we could not exist here on Earth; life would have been burnt to toast before it ever got started. Instead individual photons are emitted in the core, reabsorbed by the star's material a bit further out where the temperature is lower, re-emitted at a longer wavelength, and so on many times before eventually reaching the outer layers, or atmosphere, of the star. Only then, and at wavelengths that are much more conducive to life, do they make their way across interplanetary space. This process leads to what is called "stellar opacity" and the chemical and physical conditions of the star, layer by layer, determine how that opacity affects its overall evolution.

For the small stars in the early universe, with very few heavy elements in them, one of the main contributors to the opacity of the atmosphere was the negative Hydrogen anion, H^-, as well as free electrons. Balancing these free electrons, were positive cations: protons, H^+, Helium-Hydrogen ions, and our chemical guide, H_3^+. All of them played their role. For stars of about 40% the mass of the Sun, H_3^+ shortens the star's life. It stops the central region from being properly mixed up, so the amount of Hydrogen to be burnt to form Helium at the very center is limited. With no heat generated at the center, the star collapses prematurely, until Helium can burn to form Carbon, keeping it alive for a brief period. For even smaller stars, those only about one quarter as massive as the Sun, our guide plays another role; H_3^+ helps to obscure the radiation leaking from the center of the star out into space, slowing down its evolution and making it live longer than expected. The hunt is on for these small, old stars – in them, our guide may have something more to tell us about how the early universe evolved.

Interlude: How Our Guide Is Hooked, Lost and Caught Again

Joseph John Thomson can rightly be called the father of particle physics. Born in Cheetham Hill, a suburb of Manchester, England, a week before Christmas 1856, he went to the local university and then on to Cambridge. As Professor of Experimental Physics at the Cavendish Laboratory there, he became interested in the way in which electricity could be passed through a gas, work that owed its origin to Michael Faraday, the brilliant and original Professor of Chemistry at the Royal Institution. Thomson worked with a then relatively new piece of apparatus called the Crookes tube (later the cathode ray tube), whose use had been pioneered by British physicist William Crookes.

The Crookes, or discharge, tube was basically a tube of glass filled with gas at very low pressure with metal electrodes – an anode and a cathode – at either end. When a current was passed through the rarified gas in the tube, strange rays flowed towards the positively charged anode from the negatively charged cathode. Crookes had shown that these so-called cathode rays traveled in straight lines, could be absorbed by a metal obstacle and caused specially treated glass to phosphoresce when they hit it. But what *were* these mysterious rays?

Thomson's genius was to show that these cathode rays could be bent by a magnetic field. If the cathode rays were light waves they would not be deflected in this way. Thomson had proved that they were electrically charged, and since they traveled away from the negative cathode towards the positive anode, they had, of necessity, to be charged negatively. Not only that, Thomson showed that the amount by which the rays were bent by a given magnetic field enabled him to work out a critical ratio, or the amount of electric charge per unit mass of the component that made up the cathode ray.

Giving the regular Friday Lecture on April 30, 1897, at the Royal Institution (Britain's first public scientific laboratory), Thomson announced that he had discovered the electron, the basic element of electric current. Thomson estimated that the electron was about 1,000 times less massive than a Hydrogen atom, and that it represented a state of matter that was "the substance from which all chemical elements are built up." (We now know that Thomson's estimate for the mass of the electron was too large by nearly a factor of 2, but still remarkably accurate for the very first measurements.) The age of sub-atomic particles had begun.

If the Crookes tube that Thomson used created, or rather liberated, negatively charged electrons from an electrically neutral gas, then something positive had also to be "left behind" to maintain the overall neutrality of the gas. Back in Cambridge, Thomson had turned his interest (among many other things) to the positive left-behinds in the Crookes tube. Thomson's former student, Ernest Rutherford, had recently demonstrated that atoms were composed of a heavy, dense nucleus, surrounded by a cloud composed of Thomson's electrons. It had also become clear that molecules, or chemical arrangements of atoms, relied on electrons to hold them together. So the positive left-behinds in Thomson's Crookes tube were either atoms that had lost one or more of their electrons, or molecules that had suffered the same fate. Thomson's work in this area reached its most productive in 1911, when he discovered something that challenged chemical orthodoxy.

Thomson first published his unexpected results in two papers in the *Philosophical Magazine*, in 1911 and 1912. In the first paper he produced a table with what he called "Electrical atomic weights" listed with values from 1 to 800. Next to electrical atomic weight 3, he remarked: "More likely to be $(H_3)_{\pm}$ than $C_{\pm\pm\pm\pm}$." In the second paper he was more positive, and challenging: "There can be little doubt that this is H_3." But, he warned: "... it is not possible to reconcile its existence with ordinary conceptions about valency." Ordinary ideas of how atoms joined together to form molecules could not explain what Thomson thought he had found – it was time to reach wider audiences with his exciting new results.

On January 17, 1913, Thomson was back at the Royal Institution to give another public lecture in its prestigious Friday evening discourse series, a series that was initiated by Michael Faraday as part of his efforts to reach out to the general public. Thomson's audience for that occasion would have included non-scientific members of the middle classes, and even some London workers who were trying to improve themselves through education. Speaking of his work on the positively charged particles, he said,

"Another application of the method I should like to bring before you is the use of it for the discovery and investigation of a new substance ... sometimes there appeared on the plates a line corresponding to a particle with an atomic weight 3; this must either be a new element or a polymeric modification of hydrogen represented by H_3." Our chemical guide had made its public debut! Duly impressed, the world-leading scientific journal *Nature* carried extensive extracts of the January 17 lecture in its May 29th edition.

Thomson then summarized his work for the wider scientific community in a major lecture to the Royal Society of London entitled "Rays of Positive Electricity" on May 22, 1913. Thomson's Bakerian Lecture, as this prestigious address was called, brought together and amplified his two previous *Philosophical Magazine* papers and public talk to the Royal Institution earlier that year. It was a wide-ranging survey of work that started with the observation in 1886 by Goldstein that streams particles could be seen passing through holes in the negatively charged cathode of a discharge tube, causing the gas beyond the cathode to light up with colors that were typical of its chemical composition – "with air the light is yellowish, with hydrogen rose color, with neon the gorgeous neon red".

Some years later, it had been shown that these ray-producing streams of particles could be deflected, showing that they were positively charged. Thomson explained that very strong magnets were needed to produce the deflection, much stronger than were needed to deflect his negative electrons, proving that these positively charged particles were much heavier than electrons. Thomson himself then showed that the amount one of these positively charged particles was deflected depended on the mass of the particle, how many electric charges it carried and how fast it was moving along the discharge tube. Thomson's discharge tubes required very low gas pressures, and tens of thousands of volts to produce the particle he observed; very challenging experiments indeed. But he was a master of experimental techniques- or at least his laboratory assistant Francis Aston was!

By subjecting the particles he produced to a combination of strong magnetic and electric fields, Thomson was able to measure the speed of the particle separately from the ratio between its mass divided by its electric charge, or what he called "Electrical atomic weight." (The same year that Thomson gave his Bakerian Lecture, Robert Millikan published the first reliable measurement of the fundamental unit of electric charge, the charge on the electron.) The charged particles left tracks on a photographic plate that were curved, and Thomson was able to work out the ratio between the electric charge and the mass of the particle by exactly where it hit the photographic plate and how curved its track was.

A mass over charge ratio of 1, deduced from how curved the track was, indicated the lightest element, atomic Hydrogen H with a mass (in atomic units) of 1 and singly charged having lost its only electron. If the track indicated the ratio was 8, it was most probably Oxygen, with a mass of 16 atomic units and a double positive charge, having lost two negative electrons; but it could have been singly charged Beryllium, since that has a mass of 8. Then, Thomson used the exact position, rather than the curvature, of the track on the photographic plate to distinguish between the two. Thomson was therefore able to use his positive ray experiments as a means of chemical analysis – the foundation of modern-day mass spectroscopy.

Thomson was clearly delighted with his new instrument and his techniques for using it. He could produce very highly charged atoms; his record was achieved using atoms of Mercury, from which he could remove up to eight electrons. He even managed to get some of the freed-up electrons to combine with atoms and molecules, creating, for example, negatively charged anions of Hydrogen (H^-) and, he thought, negative Oxygen and Carbon molecules.

Thomson was about halfway through his talk when he came to the substance with a mass over charge ratio of 3, which he called X_3. The technique he had used was to focus cathode rays, or electrons, onto a metal anode, and to drive off the small amounts of gas that stick to the surface of metals. These gases were mainly Carbon Dioxide (CO_2), the gas much in the news today for its greenhouse properties that are leading to global climate change, and Hydrogen. Thomson's first idea was that he was creating Carbon atoms that had lost four electrons and become four-times positively charged. Since Carbon was known to be 12 times heavier than Hydrogen, it had a mass of 12 atomic units, and 12 atomic units divided by 4 charges would give the mass over charge ratio of 3.

But Thomson could not get Carbon to fit the bill. For a start, the track in the photographic plate was all wrong; it did not correspond to an atom with many charges, but with just 1 positive charge. Yes, Thomson could make Carbon with *two* charges, but that gave a mass over charge ratio of 6, twice the value of 3 he was finding. Moreover, if he deliberately put Carbon-containing gases such as Methane into his apparatus, he found he could not produce the elusive X_3. X_3 turned out to be remarkably persistent, and to resist various chemical treatments that would remove Carbon compounds. Thus Carbon in all its guises (and it has many) was definitely out. Could it be a new element, so far undiscovered?

Chemists had long been using Mendeleev's Periodic Table, which lists chemical elements by their mass, compared with that of Hydrogen, in such a way that they fall into chemical families that had similar properties, to predict new elements with new properties. (As we have seen, the reactive metals Sodium, Potassium and Rubidium fall into such a family, so do the corrosive gases Fluorine, Chlorine and Bromine.) The Periodic Table predicted that a new element, with a mass to charge ratio of 3, would be very reactive – almost a "super-Fluorine." Thomson found that his X_3 was "too lethargic to be consistent with the view that it was a super-Fluorine."

That left just Hydrogen.

If so, however, this was a form of Hydrogen that had never been seen before. Atomic Hydrogen denoted chemically by a single H, no problem; paired up H_2 molecules that was also well known and fine. But H_3 or rather H_3^+, since the mass to charge ratio was 3 to 1, indicating a positively charged molecule – well that was a brave conclusion to come to. Nonetheless, Thomson concluded: "If X_3 does not contain a new element and is not Carbon with four charges it must be triatomic hydrogen." Or was it?

Thomson's work, in which he had identified his X_3 as H_3^+, was reinforced in 1916 by Chicago chemist A.J. Dempster, who showed that if he increased the pressure of Hydrogen gas in his apparatus he could make more of the H_3^+ positive cation than the lighter atomic H^+ or molecular H_2^+. Dempster never fully explained this result, and it was to be nearly a decade before two Californian chemists, T.R. Hogness and E.G. Lunn, set down two chemical equations – equations that are of fundamental importance in understanding the Chemical Cosmos. In the first of the two reactions, molecular Hydrogen, H_2, is ionized to form the positive cation H_2^+; a negative electron is liberated to drift off towards the positive anode. The second reaction involved the newly formed cation colliding with more molecular Hydrogen:

$$H_2^+ + H_2 \rightarrow H_3^+ + H$$

A cation of molecular Hydrogen had teamed up with another neutral Hydrogen molecule to form our guide H_3^+ and to liberate an atom of Hydrogen. Clearly, the more molecular Hydrogen in the mix, the faster the reaction would go. This explained Dempster's experiments and proved that Thomson's hunch about the 3-to-1 mass-to-charge species was right. Hogness and Lunn did not know, however, just how important their formulation was to be.

For some hundreds of years, chemists had an idea that we now call "valency." This is ability of a chemical species to combine with others. Valency can be thought of like people holding hands, or sometimes even like octopi holding hands, depending on the chemical element in question. Some atoms seem happy to join with just 1 hand to another atom's hand, like Sodium and Chlorine in common salt. Some seem to need two or more, like Oxygen and Carbon. Hydrogen was known to be a single-handed guy – or monovalent.

So how could you form *triatomic* Hydrogen? After all, if two Hydrogen atoms each offered a hand to the third, that third Hydrogen would have to have two free hands to hold all three atoms together; and it could even be worse than that. To form an equitable combination, why should just one of the Hydrogen's have to offer two hands when the other two were only offering one apiece? That just was not fair; a genuine partnership would have all three Hydrogens offering two hands apiece to form an even-handed triangle. So to make H_3^+ at least one Hydrogen atom had to be double-handed – or bivalent – and maybe all three, and that went against what chemists at the time knew about the valency of Hydrogen.

The first person to try seriously to explain this was Niels Bohr himself, making use of his ideas of electrons existing in atoms in stable, quantized orbits. In 1918, he tried to apply his theory not just to the Hydrogen atom but to various possible Hydrogen molecules. In doing so, he introduced a very fundamental idea into the way scientists should carry out these calculations: he considered systems in which "the nuclei are at rest and the electrons move". Bohr demonstrated that the normal H_2 molecule was stable, as was triatomic Hydrogen, H_3. Unfortunately for our adventurous chemical guide, Bohr's calculations – which were based the Hydrogen atoms arranged in a straight line and the electrons moving in circular orbits – indicated that taking a negative electron away from H_3 to form the positive cation H_3^+ produced an *un*stable molecule, one that would rapidly break down.

Bohr, however, was working with Old Quantum Mechanics, as it became known. This was a hybrid beast, like a mule. In Old Quantum Mechanics, everything behaved traditionally, classically, except that a few quantum proposals were added in to explain the radiation of light by hot bodies, the photoelectron effect, the stability of atoms, and how they produced lines in their spectra. And, like a mule, Old Quantum Mechanics turned out to be fairly sterile so Bohr's early efforts really led nowhere fast.

Columbia University chemist Harold Urey's interests eventually extended to include experiments to make organic molecules, the building blocks of life, from essential inorganic starting materials, which we will look at later in Chap. 8. In 1931, however, he was working on a fundamental physical problem. Some years earlier there had been a prediction that there ought to be isotopes of Hydrogen. Since ordinary Hydrogen had the simplest and lightest nucleus possible – just one proton with an atomic mass of 1 – anything else had to be heavier. Urey was able to identify a new isotope of Hydrogen that had a mass of 2.

Urey called it Deuterium, after the Greek word for the number 2. Explaining Deuterium, a year before James Chadwick announced the identity of the neutron, was tricky; the discovery was even shocking to some. But there it was – Hydrogen with a mass of 2, and its own chemical symbol D. (Hydrogen is the only instance of chemists assigning different chemical symbols to the different isotopes. As well as Hydrogen-2 being given the symbol D, Hydrogen-3 – Tritium – is given the symbol T. But all three isotopes could be represented, respectively, as Hydrogen $_1^1H$, Deuterium $_1^2H$, and Tritium $_1^3H$.)

Deuterium caused Thomson to doubt his 1911 discovery. You could get a mass-to-charge ratio of 3 from a molecule made up of just one atom of normal Hydrogen, with mass 1, and one atom of Deuterium, with mass 2, making HD^+, which would be analogous to the cation of normal molecular Hydrogen gas, H_2^+. Producing a cation of Deuterated Hydrogen gas, as it was called, was also easier to understand chemically than the triatomic Hydrogen ion that Thomson first thought he had produced: "The evidence seems to me to leave little doubt that the gas I called H_3^+ more than 20 years ago is the same as that which is now called heavy hydrogen," he said.

Exit our guide, H_3^+, with a heavy Hydrogen heart, abandoned by its discoverer.

But not quite; Thomson at least still felt that H_3^+ accounted for the *transient* mass-to-charge-3 positive rays he had measured, even if not the longer lived ones. Also in the 20 years since Thomson's initial work, many other scientists had taken the field, both experimentally and theoretically, on behalf of our chemical guide. Mass-to-charge-3 positive particles showed up regularly in experiments, so there was little doubt it existed. Despite Bohr's early attempts, what remained more difficult to understand was *how* H_3^+ existed.

This lack of understanding was to the great theoretical chemist Henry Eyring "*the* scandal of modern chemistry." After all, if you could not understand the behavior of a chemical as simple as Hydrogen, what hope was there for more complicated substances? Eyring, who worked at the prestigious Institute for Advanced Study at Princeton University along with Albert Einstein, had set himself the challenge of dealing with this "scandal of modern chemistry," and it was to occupy him and his colleague, Joseph Hirschfelder, for two intense years between 1936 and 1938. In a series of papers, they tried to make sense of the triatomic Hydrogen family: neutral H_3, the positively charged cation H_3^+, and even the negatively charged anion, H_3^-.

To do this, they had to make use of some recent, and in some ways shocking, advances in understanding how chemical atoms bond together to make molecules. Thomson's original question about the valency of Hydrogen had been prompted by an understanding that that atoms bonded together such that a *pair* of

Shooting the Rapids: The Life and Death of the Earliest Stars 53

atoms shared the bond (or bonds) between them. Hirschfelder and Eyring dropped this assumption and carried out calculations for an arrangement of three Hydrogen atoms in which all three bonded simultaneously to make a structure that had the three Hydrogens arranged in a line. Their calculations were based on the then relatively new theories of modern Quantum Mechanics.

The peaks and troughs associated with a wave in the ocean can be described by a simple mathematical formula – usually a sine or cosine function. That is the ocean wave's wave-function; it describes the state of the ocean. The island of Hawaii gets warnings about the state of the Pacific Ocean in terms of the height of the waves. Some breakers can reach 10 m and more, and only the most experienced surfers will brave them. The bay front downtown area of Hilo was twice hit by devastating tsunamis in 1946 and 1960; the 1946 tidal wave was over 12 m in height. Therefore if you want to know the state of the ocean, measure the height of the waves.

While Einstein had shown that light waves also acted as particles, French scientist Louis de Broglie soon demonstrated that a moving electron, or any other atomic or sub-atomic particle, would behave not just as a particle, but also as a wave; that electron would have a wavelength given by h divided by the momentum (the mass times the speed) of the particle. So there was a nice symmetry in Nature: light, long considered to be a wave, also acts like a particle; electrons and other atomic and sub-atomic particles, under certain circumstances, show a wavelike behavior.

This wave-particle duality turns out to be the very foundation of Quantum Mechanics. The electron in an atom or molecule is in a *state* that is described by its wave function. If you want to know the state of the electron – where it is, how fast it is moving, how energetic it is – you have to measure it by carrying out an operation on the wave function. In Quantum Mechanics, wave functions describe state and operators enable you to measure a particular property of that state. To understand our chemical guide H_3^+, full-blown Quantum Mechanics must be applied to the problem, not the hybrid mule that Niels Bohr was using in 1918.

Bohr did, nonetheless, point the way. His notion that the three Hydrogen atomic nuclei could be considered as fixed while the electrons moved was based on the sound physical observation that the Hydrogen nucleus (the proton) is more than 1,800 times heavier than the electron. So the situation of electrons and atomic nuclei in a molecule is a bit like a swarm of flies around a herd of cows. The buzzing flies (electrons) do not cause the cows (nuclei) to move, but as the cows move off to new pasture, the swarm of flies instantaneously sets off with them. Movements of the electron around the nucleus do not cause it to move appreciably, no more than the orbiting planets of our Solar System cause the Sun to move – much. As the nuclei move, however, the electrons immediately adjust their positions to follow them.

In the new Quantum Mechanics, this observation, this separation of electron motion from nuclear motion became enshrined as the Born-Oppenheimer Approximation, after Max Born and Robert Oppenheimer who first investigated how good an approximation it was. It was a simplification that eventually was to make it possible to calculate the state of multi-atom (polyatomic) molecules with extremely high accuracy, such that for simple molecules measured, laboratory spectra and calculated line wavelengths would match to an accuracy greater than the actual lab measurements themselves.

The Born-Oppenheimer approximation gives rise to a simplified picture of a molecule in which the atomic nuclei vibrate and rotate on a "potential energy surface" that has been generated by calculating how much energy the electrons will have when the atomic nuclei in the molecule are held fixed at different positions. Potential energy is the energy anything has simply as a result of where it is; a china cup placed on a table has potential energy because it has been lifted above floor level. Knock the cup off the table and it will probably smash as it hits the floor. The potential energy, therefore, of being on the table is first converted into kinetic energy, due to the cup falling under the force of Earth's gravity, which in turn generates enough energy to break the cup apart as it and the floor collide. Some of the energy will even have been converted into sound, although the curses that usually accompany a cup smashing come from a different source.

It is relatively easy to visualize the potential energy surface for a molecule that is made up of just two atoms, such as diatomic molecular Hydrogen, H_2. A graph of the potential energy plotted against the distance between the two Hydrogen atoms making up the molecule has the shape of a valley. On the nearside – when the two atoms are close together – a potential energy cliff rises almost sheer to a great height. On the far side, there is an energy hill to climb, but it is not as high as the cliff because it plateaus out. The two individual, uncombined, free Hydrogen atoms can be represented as the plateau, far away from the valley.

Moving across the plateau towards the valley is like bringing the two Hydrogen atoms towards one another, until a chemical bond attracting them to one another begins to take over. Then the atoms are brought closer together in the molecular Hydrogen form, H_2. The bottom of the valley represents the most stable distance between the two Hydrogen atoms in the molecule – it is the point of lowest potential energy, and the deeper the valley the more stable the molecule. Bringing the Hydrogen atoms even closer together, however, means that forces of repulsion between the two negatively charged electrons start to come strongly into play. It is like forcing the molecule to climb the steep energy cliff on the nearside of the valley; the potential energy increases steeply for every small decrease in the distance between the atoms.

The electronic potential energy surface tells us about the stability of a molecule and what shape makes it most stable. Just as in individual atoms, molecules can have a set of electronic states and transitions in which the molecule jumps from one electronic potential energy surface to another level, give rise to spectra. Although we can represent the potential energies as a function of the distances between atoms in more complicated molecules as a potential energy surface, the picture is more complex than for a simple two-atom molecule like H_2. Just to increase the number of atoms to 3 involves many more calculations. For now there are three distances between atoms to consider – between atom 1 and atom 2, between atom 2 and atom 3, and between atom 3 and atom 1. As the number of atoms goes up, the problem of understanding the molecule and of calculating its potential energy surface becomes more and more daunting.

One of the ideas that came out of Quantum Mechanics was that the real state of an atom or molecule could be made up of combinations of simpler states. In the case of H_3 and H_3^+, in some instances the bond was between the Hydrogen atoms at either end of the molecule; in others the middle Hydrogen bonded alternately with the atom on either side of it. But the overall structure of triatomic Hydrogen was, according to Quantum Mechanics, simultaneously a *superposition* of these simpler structures.

Using this approach in 1936, Hirschfelder and Eyring found that H_3, neutral triatomic Hydrogen, was stable compared with its component parts, three Hydrogen atoms, or a normal diatomic Hydrogen molecule and a Hydrogen atom. The triatomic Hydrogen ion, our adventurous guide H_3^+, was even more stable by a factor of more than 2½ – a surprise, since taking an electron away from a molecule to make it ionized usually resulted in making it less stable, not more so.

Over the next couple of years Hirschfelder, worked on more and more sophisticated calculations of the energy and stability of triatomic Hydrogen species. By 1938, he had moved to the Chemistry Department at the University of Wisconsin. He had also taken the bold step of calculating what would happen if he allowed for the molecules to become bent.

Hirschfelder's calculations, all carried out before the invention of the computer, showed that the most stable form of the neutral H_3 was still to have the three Hydrogen's line up alongside one another. That was not the case for H_3^+ though; Hirschfelder found that a triangular form of the molecule was the most stable, almost three times as stable as his original calculation of the stability of the parent, neutral H_3. Our adventurous chemical guide had now really arrived on the scene – in discharge tubes and in the challenging calculations of the theoretical chemist. No more would it be disowned by its original discoverer, confused with heavy Hydrogen, Deuterium.

Hirschfelder's work in 1938 went beyond simply showing how stable H_3^+ was and demonstrating that it was most stable as a triangle. He was also able to calculate that, as a bit more energy was pumped into the molecule, it would vibrate. "The vibration frequencies of H_3^+ are estimated but their exact value[s] … are somewhat in doubt. Two of the frequencies should be infrared active and susceptible to direct experimental measurement," he predicted.

Sort out one chemical scandal, set another in motion. For *the* scandal, as Eyring would have it, from 1938 onwards was that no one could now meet Hirschfelder's challenge and measure the spectrum of H_3^+. Hirschfelder's challenge was not to find spectral lines of H_3^+ produced by changes in electron orbits; he was after a different kind of molecular change.

Unlike the atoms of which they are made, molecules are extended objects. The atomic nuclei that go into making up a molecule are distributed in space with respect to each other, albeit that the separation between these nuclei are of the order of tenths of a nanometer. The distance between the nuclei can change rapidly when the molecule vibrates; they tumble in space and the molecule rotates. And, as you might expect from Quantum Mechanics, these vibrations and rotations are quantized. The molecule has to be associated with a vibrational and a rotational energy state that can only have a definite value, even if these energies are usually much less than those associated with the electronic states. Not all values of vibrational and rotational energy are possible – if you cannot get change of an electronic dollar, you certainly cannot get change of a vibration-rotation cent. Hirschfelder's challenge then was to measure the spectrum of H_3^+ due to a change in the vibrational state of the molecule.

Given the simple nature of H_3^+, it should have been straightforward. In the 1930s, spectroscopists were routinely measuring the spectrum of much more complicated molecules than the triatomic Hydrogen ion. But Hirschfelder's challenge was to endure, unanswered, for four decades. With the onset of World War II, the efforts of much of the scientific community were directed to military tasks, and it was not for a decade after Hirschfelder's work that anyone seems to have taken up the challenge of H_3^+ again. Additionally it took until the mid-1960s, and the widespread introduction of powerful computers into academia, before the next important step in understanding H_3^+ could be taken.

The H_3^+ of Hirschfelder was a bent-out-of-shape affair. It was not a right-angled triangle, of the sort beloved of Pythagoras, and it was not a very symmetric triangle, in which all sides were the same length (an equilateral triangle). His 1938 result was, however, based on the idea that the bonds that hold a molecule together are composed by pairs of atoms sharing electrons – two-center bonds.

Hirschfelder found that calculations that assumed the electrons were "delocalized," shared equally across all three atoms of the entire H_3^+ molecule, produced a less stable molecule. Earlier, the English chemist Charles Coulson had suggested that the ion would be more stable if all the bonds and angles were the same, that H_3^+ was an *equilateral* triangle. But he had not been able to calculate the energy sufficiently accurately to demonstrate that this really was the case.

Accurate calculations of the delocalized electronic orbitals required solving a problem involving many bodies – a job for computers, but when these became standard in the 1960s, the correct shape was produced. Our chemical guide was now a triangle with all three sides the same length, as Coulson had suggested. What is more, two groups, one led by Ralph Christofferson, the other by Harold Conroy, nearly simultaneously produced the same result, convincing other scientists, too, that H_3^+ really was an equilateral triangle – at least in theory.

The final clincher that our chemical guide had a nice symmetric triangular shape came in the form of a brutal experiment. Particle physicists love accelerators, large pieces of equipment that make use of giant electromagnets to fire particles up to very high velocities. In the largest, such as at the European nuclear research center, CERN, protons can be accelerated close to the speed of light and smashed into one another to produce strange forms of matter. Our adventurer did not get quite that treatment, but close enough.

As so often in the adventures of our chemical guide, a truly international team was assembled. The Argonne National Laboratory houses ATLAS, or the Argonne Tandem Linac Accelerator System. ATLAS was the first facility in the world to use superconducting magnets to accelerate particles heavier than electrons. ATLAS is ideally suited to accelerate positive cations, which are much heavier than electrons. ATLAS produced its first beam of accelerated ions in 1978, and part of the input to ATLAS was its Dynamitron injector. In this equipment, H_3^+ ions were being accelerated up to 11,500 km/s, nearly 4% of the speed of light. Half a world away, at the Institut de Physique Nucléaire in Lyon, France, a similar experiment was being performed.

These poor high-velocity H_3^+ projectiles were being fired at film of Carbon less than 100 atoms thick, around 10 nm. As the H_3^+ ions crashed through the Carbon film, all their electrons were stripped away, leaving three Hydrogen nuclei – protons – in close proximity to one another. Three closely associated positively charged protons repelled each other very violently, and the resulting "Coulomb explosion" sent the individual particles flying away from one another. At the detectors, a number of proton peaks were detected, and at Argonne and Lyon, the story was the same – only if the H_3^+ molecule were an equilateral triangle, could the observed patterns be explained.

Two independent experiments obtaining the same answer, agreeing closely with theoretical predictions; if that were not enough, at the Weizmann Institute in Rehovot, Israel, a third team was working on the same type of experiment. This time, however, they did not measure peaks of protons arriving at a detector. Instead they took a photograph of the Hydrogen nuclei as they came flying out of the thin Carbon film, denuded of their electrons. Triangles showed up everywhere, averaging to an equilateral one. This, contrary to the rules of baseball, was three strikes- and you're in!

Gerhard Herzberg won the Nobel Prize for a lifetime's service to spectroscopy in 1971, at the age of 67. But lifetime? Herzberg was just getting started! In 1975, he became Distinguished Service Scientist of the institute in Ottawa, Canada that was to bear his name. When he "officially" retired in 1994, Herzberg was still taking a lively interest in spectroscopy at a special conference organized to celebrate his 90th birthday, attending all the sessions and bombarding the speakers with penetrating questions. The Herzberg Institute rapidly became one of the key centers for molecular spectroscopy, producing world-leading scientists as if off a conveyor belt.

One of the most inventive of the Herzberg's "products" is Takeshi Oka, who we first met in Chap. 1. Oka was born in Tokyo in 1932 and later studied at the University, obtaining his doctorate in 1960. But Oka – he does not allow even close friends and colleagues to use his first name – found the atmosphere in Japanese academia too stiff and formal for his relaxed personality. (Years later and as a senior and world-acclaimed professor, Oka much preferred to eat pizza in the lab with his students than to go to formal conference dinners.)

After three further years as a research fellow in Tokyo, Oka left Japan for the Herzberg Institute in 1963, where he would remain for 18 highly productive years. During that time, he published 80 scientific papers on molecular physics, carrying out detailed measurements and working on the theory of how to interpret what he observed. Under the leadership of Herzberg, Oka began work that would play a major contribution to what he calls "the ongoing unification of chemistry and astronomy", or astrochemistry, the foundation of our understanding of the Chemical Cosmos.

Herzberg, among others, had drawn attention to the importance that a molecular ion such as H_3^+ could play in space; he had tried hard to measure its spectrum in case it could be detected there. After all, 90% of all atoms in the universe were Hydrogen, and much of this was in the form of paired up molecular Hydrogen, H_2, in vast clouds of dust and gas that lay between the stars, and out of which stars themselves form. Ionizing radiation – ultraviolet light and energetic particles – poured through the clouds, and it was inevitable that, with so much Hydrogen around, H_3^+ would be formed. First, however, one had to know what one was looking for.

This is where Oka made the key discovery of his career, at least as far as the world of astrophysics was concerned. Two new technical developments helped him in his work: the first was the development of discharge tubes in which it was possible to produce a very long column of positive cations. Although this had first been applied to measuring microwave spectra, Oka saw no reason why it could not be applied to search for the, much shorter wavelength, infrared spectrum of H_3^+ that Hirschfelder had suggested should be detectable.

The second development was in the field of lasers. The laser is now ubiquitous, with applications as diverse as dissecting British secret agents, eye surgery and playing CDs and DVDs. It had been "invented" at Bell Labs by Charles Townes, Arthur Schawlow and Gordon Gould, and the first laser to work with optical light was a ruby laser made by Theodore Maiman in 1960. (The history of the laser is rife with disagreement and lawsuits, so I have covered all of the bases here.)

The laser gave very intense beams of light, much more powerful than the traditional glows used by spectroscopists, and they were quick to see its potential. What Oka was able to exploit was the fact that lasers could now be made to produce intense infrared light beams. This was achieved by making use of wavelength differences between a laser filled with Argon gas and a laser in which made use of organic dyes, such as Rhodamine.

Argon lasers emit green light with a wavelength of 514.5 nm. Organic dyes could be made that would laze over a wide range of wavelengths from around 550–700 nm. By mixing the beams of Argon and dye lasers, it was possible to produce powerful beams of light at the wavelengths – in the wavelength range of thousands of nanometers, or microns – which molecules would absorb or emit.

Oka successfully brought the two developments of long columns of positive cations and infrared lasers together at the Herzberg Institute to look for the spectrum of H_3^+. "The first spectral line of H_3^+ was found on April 25, 1980," he reported. "By May 9, 10 lines had been observed and the approximate spectral pattern and their response to varying discharge condition suggested strongly that indeed the v_2 band of H_3^+ had been obtained." But the lines had not been "assigned" and had not been classified according to the laws of Quantum Mechanics; they had not been properly labeled their "quantum numbers." And that was a must if other spectroscopists were to believe Oka's work.

Oka knew that the lines he had measured did belong to the first vibrational band of H_3^+. This "v_2 band" was produced by H_3^+ changing its vibrational state by one vibrational quantum. In Oka's discharge tube, H_3^+ was absorbing infrared radiation as it jumped from its ground state to its first vibrationally excited state. Oka had met Hirschfelder's challenge set more than 40 years previously; the vibration was, indeed, infrared active, and infrared lasers had made it light up.

Now needed was how much rotation was associated with the lines Oka had measured, as the H_3^+ molecules tumbled through space. The first line he had detected had a precise wavelength of 3.668533 microns (a micron is one millionth of a meter), an indication of how precise the measurements were. Others were scattered throughout the wavelength range from 3 to 4 μm. The news that Oka had discovered the elusive H_3^+ spectrum set the Herzberg Institute abuzz with excitement.

Normally, spectroscopists can assign their lines because they can see a regular pattern in them. Some lines represent molecules that have picked up a quantum of rotation, some have lost one, and others have stayed with the same rotational energy. Those that pick up rotational energy as they jump up a vibrational state absorb radiation at shorter wavelengths; those that lose rotational energy at longer. Those that do not change rotational state ought to form a cluster in the middle, but Oka's spectrum was nothing as neat as that so assigning the spectrum would be challenging. Fortunately, also working at the Herzberg Institute in 1980 was theoretical chemist Jim Watson. Watson had been developing computer programs precisely for the purpose of assigning difficult spectra. The night that Oka measured his tenth line, Watson had enough to get to work. His program fitted the lines perfectly.

There was now no room for doubt – Oka had really measured the H_3^+ spectrum to the satisfaction of the world's spectroscopists, and Watson had assigned it to a series of rotational and vibrational quantum numbers. Our guide had truly arrived and been fingerprinted at the border control of chemistry. So where now for our adventurous molecule? We take a look back out into space, to the giant gas clouds that fill our galaxy and others, to what is known as the Interstellar Medium, the broad river of the Chemical Cosmos.

4. Heading Downstream and Cooking by Starlight

The visible night sky is full of bright stars. Viewed from very dark locations, literally thousands of them are visible, and it is easy to pick out the Milky Way in the sky. This band of stars, so many that they make up a continuous milky glow, is our view of the main part of the galaxy in which we reside. Astronomers measure the size of the Milky Way in units of light years: light covers nearly 9½ million, million kilometers every year at a rate of 300,000 km every second. From its center to its outside edge, our galaxy, the Milky Way, measures roughly 50,000 light years. It is a spiral galaxy, with arms spiraling out from the center like streamers from a spinning dancer.

As well as the millions of stars in the Milky Way, those with good eyesight can also make out dark "lanes" where stars seem to be almost entirely missing. They are not, but vast clouds of dust and gas blot out their starlight at visible wavelengths. These clouds make up more than half of the normal matter in the plane of the Milky Way. The gas density in them is vanishing small, on average 10 billion, billion times less than the air we breathe, just a few million atoms or molecules per cubic meter. (The air we are breathing while reading this book contains around 25 to 30 million, billion, billion molecules in every cubic meter. Every breath we take contains about 15,000, billion, billion molecules.) Even the densest clouds probably have no more than a few million, million molecules per cubic meter. These interstellar clouds make up for that low density by their size; light can take years and decades to cross them, and – even at the low densities of the gas and dust – the photons of light often do not make it through before they are absorbed. That is why the lanes in the Milky Way look so dark (Figure 4.1).

FIGURE 4.1 A view of our Milky Way galaxy, showing "lanes" due to gas and dust clouds: *credit – NASA*.

The gas clouds of the Milky Way are chemically rich. Their alphabet soup of chemical elements have been cycled and recycled through generation after generation of stars, some explosive giants that ended as supernovae, others more modest creatures like our own Sun. Together, the hot and cold gas clouds lying between the stars are termed the Interstellar Medium, or ISM. The ISM is where our chemical guide H_3^+ is most at home, so let's take the tour down the main river of cosmic chemistry.

The name "Interstellar Medium" hardly does justice to this all-pervasive matter. For a start, the term sounds like a mystic claiming to get in touch not just with our departed loved ones, but with those from worlds throughout the galaxy. Secondly, the interstellar environments covered by this one term, whilst tenuous, are varied. Take our own neighborhood, for example.

Our Sun, the Earth and the rest of the Solar System occupy a fairly typical (if such things exist) location in the Milky Way galaxy, not too up market, but not in the slums, either. Here we are, not quite two thirds of the way out from the center of the galaxy. The bright light of downtown Milky Way takes about 28,000 years to reach us; therefore we are 28,000 light years from the center of the galaxy, over 250 million billion kilometers out in the suburbs. On the Orion arm of our spiral galaxy, as we looking outwards towards intergalactic space, we see the constellation of Orion the Hunter; looking inwards towards the center of the galaxy, we see the constellation of Sagittarius the Bowman.

The Solar System as a whole is surrounded by a warm tenuous gas made up of Hydrogen atoms, coexisting with extremely hot plasma of protons and electrons. Together these are known as the Local Bubble, extending some 300 light years across. Even within one region, there can be two flavors of ISM, like hot chocolate sauce on cold vanilla ice cream. Gas densities in the Local Bubble are probably less than 100,000 in every cubic meter; ten times lower than the average ISM. This relative clearing in the ISM is shaped more like an hourglass than a bubble, evidence of giant supernova shocks driving away most of the gas and dust.

The Local Bubble is bounded by a denser ISM, where gas densities are between 100,000 and 30 million in every cubic meter, still low enough for this to be called the *Diffuse* Interstellar Medium. Our immediate neighbors, however, also include dense gas clouds in the constellation of Taurus the Bull and Aquila the Eagle, where temperatures are very cold and gas densities are measured in billions and tens of billions in every cubic meter. These clouds are known as the *Dense* Interstellar Medium. Almost directly opposite Taurus are the remains of a supernova explosion that happened in the southern constellation of Vela the Ship Sail around the time our ancestors were taking up farming, some 12,000 years ago. And at one end of the Local Bubble is Orion, more than 400 light years distant from Earth.

Orion is one of the most easily recognizable constellations in the northern hemisphere. Even in the most light-polluted cities of the northern hemisphere, the red giant star Betelgeuse, the three belt stars and the bright hot Rigel are easy to make out. Behind these bright stars, the entire constellation is a mass of gas and dust clouds. Some of these clouds are very hot: the bright Orion Nebula that forms the center of the Hunter's sword has gas temperatures that reach up to 10,000K. Light takes some 24 years to cross the

FIGURE 4.2 The Orion Nebula shows trails due to "bullets" of rapidly moving gas passing through it: *credit – Gemini Observatory*.

full 200 million million kilometers of the Nebula. In the center of the Nebula, bright young stars are forming from the denser knots of gas and dust created when giant shock waves pass through it. These stars drive winds of tens of kilometers a second, bullets of hot gas through the swirling clouds, creating shock waves of their own.

The Orion Nebula is seared by intense starlight from some 2,000 stars of all ages. Its clouds radiate a warm pink glow, the alpha line of atomic Hydrogen at 656.3 nm. The presence of this H-alpha emission is testimony to the fact that Hydrogen atoms are in abundance, that they have been ionized to free Hydrogen nuclei and electrons, and these are recombining back into neutral atoms once more, repeating an endless cycle of ionization and recombination. These hot bright clouds attest to the energetic nature of star-formation (Figure 4.2).

Many of the clouds in Orion are cold, however, at temperatures around or below the boiling point of liquid Nitrogen, just 77° above absolute zero. The Horsehead Nebula, found just below the left star in Orion's Belt, is dark against the pink of surrounding atomic Hydrogen gas. It is a dense mass of gas and dust that not even starlight can fully penetrate, and its very darkness is a sign of the complex chemistry that goes on between the stars. Both the Orion Nebula and the Horsehead Nebula are parts of the Orion Molecular Cloud Complex, hundreds of light years across, and named for its rich chemical composition. Nor is the Orion constellation alone in being home to vast clouds of gas and dust; nor does it lay claim to host the coldest of clouds.

Some clouds, such as those in the constellation of Taurus, have even lower temperatures of just 10K or 20K. The ultraviolet and visible starlight that have fallen on the cold clouds in the ISM are absorbed and then radiated back out into space at much longer wavelengths, in the infrared, microwave and, at the lowest temperatures, at radio frequencies – by molecules. As a result, such clouds cause dark lines to appear in the ultraviolet and visible part of the spectrum of the stars that lie behind or embedded in them. These lines were the first clues that there were giant, molecule-containing, gas clouds between the stars.

In 1922, Mary Lea Heger from the Lick Observatory in California noticed that there were broad lines in the visible spectrum of stars wherever she looked (Figure 4.3). These ubiquitous features were called the Diffuse Interstellar Bands – or DIBs – but no explanation as to what exactly was causing them was forthcoming: they did not seem to be the result of atomic transitions. Could they be due to molecules? Despite serious doubts by many astronomers, confirmation that molecules could exist in the harsh environment of space was soon forthcoming. In the 1930s, astronomers working at the Mount Wilson telescope in California measured the starlight coming from a series of blue stars and found dark lines in the ultraviolet end of the spectrum due to some exotic, but simple, Carbon-bearing molecules – CH and its ionized progeny CH^+, and the highly reactive Cyanide radical CN. These first detections of molecules in interstellar space opened up a new area of astronomy called astrochemistry. As with so many other branches of science, however, World War II put much of the research effort into astrochemistry on hold.

Radio astronomy really took off in the period immediately post-World War II, and one of the earliest of the radio observatories

68   The Chemical Cosmos

FIGURE 4.3 Mary Lea Heger and her husband Donald Shane: *credit – Donald E. Osterbrock, courtesy AIP Emilio Segre Visual Archives, Physics Today Collection.*

was at Jodrell Bank, near the industrial city of Manchester, in Northwest England. Led by its energetic director, Bernard (later Sir Bernard) Lovell, it started as an astronomical site in 1945, immediately after World War II. By 1947, Lovell had built a transit radio telescope. With a diameter of just over 7 meters (m), it was at the time the largest radio telescope in the world. The transit telescope pointed directly up into the sky, hence its name, since it could only observe objects when they "transited" directly overhead.

Emission from Hydrogen atoms at a radio wavelength of 21 centimeters (cm) seemed to be ubiquitous, a key indicator of gas at a temperature of a few thousand degrees. With their huge iconic steel dishes, radio observatories captured the public imagination. After all, radio meant communication, and communication from space meant … The science fiction writers had blueprints beamed from alien intelligences to create perfect replicants, and battles for domination of Earth between galactic superpowers, slugging it out in the van Allen radiation belts. In 1954, radio waves were even detected coming from the planet Jupiter: were the aliens getting closer? For astronomers, the dark clouds that crisscrossed the Milky Way were to shine brightly with radio waves announcing the presence of literally dozens of molecules. But these were not just simple molecules, like our familiar old diatomic Hydrogen; there were complex creatures out there to be found.

Over the next decade, Lovell built the iconic Mark I (now Lovell) telescope. When it was built, this steel wire dish with a diameter of just over 76 m and a collecting area more than 100 times greater than the transit telescope was the giant of radio astronomy. Fully steerable, it did not have to wait for heavenly objects to cross its path; instead it could be turned towards them as required. In the spirit of turning swords into ploughshares, the gun turret mechanisms from HMS Revenge and Royal Sovereign were commandeered to ensure perfect pointing. On October 12, 1957, just after it was finished, the Mark I was able to detect that the Soviet Union had launched the world's first artificial satellite, Sputnik, and to track its rocket across the sky.

Among its numerous notable achievements was the study of a new phenomenon for astronomers, the maser, or the microwave equivalent of the laser. Masers (and lasers) work by pumping atoms or molecules enclosed in an electrical cavity up into excited states, then stimulating them to emit radiation all together in a way that is known as "coherent". Powerful beams of light, or microwaves, can then be generated. Charles Townes, of Columbia University, had won the 1964 Nobel Prize for Physics for co-inventing the maser-laser in the early 1950s. But in 1957, he had suggested that masers might also be found outside of the electronics laboratory, in astrophysical objects such as the atmospheres of planets, the coma of comets, in small gas clouds, or nebulae.

Townes' astrophysical masers were first detected in 1965 by Harold Weaver and his colleagues working at the Hat Creek Observatory in California, during a search of the Orion nebula and other gas clouds for the simple molecule OH, a derivative of Water. This molecule had first been detected in the spectrum of the Cassiopeia A gas cloud two years earlier. Weaver's team found unexpectedly strong emission at a wavelength of 18 cm. This was a frequency close to what they were expecting for their OH molecule but so bright that they did not dare to conclude it could be OH; instead, they called it "Mysterium". Weaver's discovery was followed up by scientists led by Rod Davies working at Jodrell Bank, using the new Mark II telescope, half the size of the Mark I, but better suited for high-frequency radio observations. The Jodrell astronomers and others deduced that the 18 cm line was indeed the OH molecule, but in an environment in which it was being pumped up into an excited state to an extent that would only occur for sources with an equivalent temperature of a billion degrees or more, much, much too hot for molecules to survive. These bright lines were also being formed in regions of intense magnetic fields. In the years that have followed, OH masers, and those of other molecules including Water, have been shown to trace highly energetic regions in gas clouds, such as those where stars are formed or where the clouds of gas thrown out by supernovae create vast shocks as they hit the ambient interstellar gas.

By 1971, Townes, now at the University of California, and his team were in a position to put together a comprehensive review of the state of astrochemistry for the American Association for the Advancement of Science (AAAS). Writing in its prestige house journal *Science*, they reported that since the first radio detection of OH molecules in the ISM in 1963, organic molecules such as Methanol, Acetic Acid – the active components of wood alcohol and vinegar, respectively – and several poisonous Cyanides had all been found through their radio fingerprint spectra. Adding to this deadly brew, the fingerprint lines of the Carbon-Oxygen compound Carbon Monoxide, CO, were now being found throughout the ISM. The signature CO line at 2.6 millimeters (mm) was a key indicator of molecule-rich, colder gas; although there was only one CO for every 10,000 Hydrogen molecules, this deadly gas is a much more powerful emitter and could clearly be seen. Water had been found the year previously. Formaldehyde, the liquid used to preserve biological specimens, and its heavier cousin, Acetaldehyde, were there. In addition to molecular Hydrogen, H_2, some 20 identifiable molecules had been found, including three Sulfur-bearing molecules, and there was evidence of others. Clearly, there was a myriad of chemical reactions going on out there. The question was *how* particularly as many of the clouds being surveyed had temperatures of only a few tens of degrees above absolute zero, some 200K below the freezing point of ice.

Anyone who has had any acquaintance with the chemistry laboratory or the kitchen has a good idea that chemical reactions usually need heat. You heat the test-tube or saucepan full of the ingredients to produce the chemicals or food that you want. To get molecules – be they from the lab bottle or the supermarket shelf – to break up and combine to form new ones with different properties and flavors from the ones you started with, you need energy. There is an "energy barrier" that has to be overcome, either by the flame of the Bunsen burner or the kitchen oven.

However in many of the giant interstellar gas clouds where numerous molecules have been found, there is very little of that energy to spare in order to get over energy barriers and start the chemistry going. What little there is generally provided by ultraviolet starlight or cosmic rays, ubiquitous high-energy electrons, protons and other atomic nuclei that have been produced, for the most part, by supernova explosions. All of these cosmic rays can ionize the gases found in the interstellar clouds, albeit slowly. Even in the denser gas clouds, ionization by cosmic rays occurs only once every 2 weeks in each typical cubic meter of gas. Nonetheless, since much of the gas is composed of diatomic Hydrogen molecules, ions composed of Hydrogen are extremely likely. In 1925, Hogness and Lunn had written down their famous chemical reaction:

$$H_2^+ + H_2 \rightarrow H_3^+ + H$$

Once the ionization of molecular Hydrogen occurred, H_3^+ would be formed very rapidly. Interstellar clouds that were rich in Hydrogen molecules provided just the right environment for our guide to form.

The first suggestion that H_3^+ might be "out there" came from a team of Atlanta physicists led by Earl McDaniel in 1961. Their laboratory experiments had confirmed A.J. Dempster's 1916 observations that H_3^+ was formed very rapidly in mixtures of ionized Hydrogen molecules and neutral diatomic Hydrogen. They also found that, once formed, it was also very stable. So, depending on just how dense the Hydrogen gas was in the interstellar clouds that had come under the scrutiny of radio astronomers, they thought it worth a shot to see if our chemical guide was around in appreciable quantities: "It now appears desirable to consider the possibilities for detecting H_3^+," they advised.

Moreover, it became clear that our guide could play a pivotal role in the kind of chemistry that was going on in the vast interstellar clouds. Shortly after Charles Townes and his team published their review of the advances made by radio astronomers in detecting molecules in space, Harvard chemists Eric Herbst and Bill Klemperer, and Bill Watson of the University of Illinois, pointed out two key properties of H_3^+ that made it just the kind of agitator that could get a long chain of chemical reactions started. Firstly, although it was certainly a stable molecule, H_3^+ was also very reactive. Chemists have a notion of "proton affinity", how anxious an atom or molecule is to grab hold of and hang on to a proton, H^+. In the chemical reaction that makes H_3^+ from H_2^+ and H_2, we can think of H_2^+ as a protonated Hydrogen atom. Along comes a diatomic Hydrogen molecule, H_2, which has a higher proton affinity than H, and the proton jumps ship to form H_3^+ leaving a lonely Hydrogen atom behind.

It turns out that this makes H_3^+ good and stable when the gas is pure Hydrogen, but pretty much everything else has a higher proton affinity than diatomic Hydrogen molecules, H_2. Put a little Carbon or Oxygen into the mix, and the proton in H_3^+ will jump ship again, producing protonated Carbon, CH^+, or protonated, Oxygen, OH^+, and return a diatomic Hydrogen molecule, H_2. No more H_3^+, but in sacrificing itself our noble guide has started a chain of chemical reactions that builds bigger and more complex molecules in the interstellar clouds. H_3^+ similarly hands over a proton to Carbon Monoxide, diatomic Nitrogen molecules, or Water. Thus it plays a vital role in starting the main chemical pathways to complex molecules containing the three essential life elements: Carbon, Nitrogen and Oxygen (Figure 4.4).

The second property of H_3^+ that Herbst, Klemperer and Watson pointed out was that the reactions of H_3^+ with Oxygen and other chemicals did not need heat to get them started. There was no energy barrier; you could cook *without* gas, or electricity, for that matter! The fact that H_3^+ is positively charged means that it is naturally attracted to the cloud of negatively charged electrons surrounding neutral atoms and molecules. Once it was up close and personal, it could then let go of its proton, allowing the

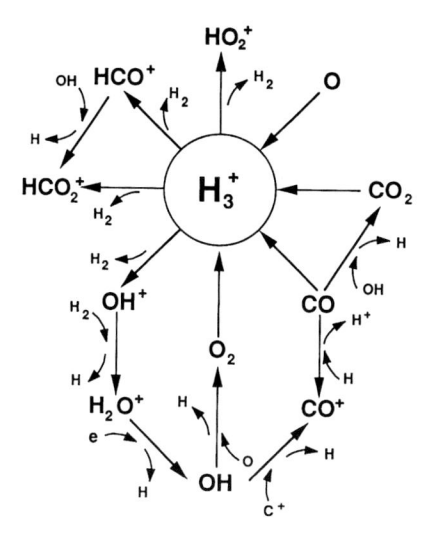

FIGURE 4.4 In the Interstellar Medium, H_3^+ initiates a web of reactions that lead on to the formation of larger molecules. This "web" involves chemistry with Carbon- and Oxygen-containing molecules: *credit – Steve Miller*.

remaining diatomic Hydrogen molecule to drift off. In the cold of dense interstellar clouds, with temperatures only a few tens of degrees above absolute zero, reactions involving ions – electrically charged molecules – go much faster than reactions in which both parties are neutral molecules. Even so, at the gas densities prevailing in interstellar clouds, it still takes an H_3^+ molecule between 4 months and 4 years to find an Oxygen atom and react with it. Astronomers, or rather the clouds that they study, have lots of time: anything from hundreds of thousands to millions of years. It is slow, but sure, cooking.

In the chemical laboratory or the kitchen, where the heat to get over the energy barrier to a chemical reaction is available at the turn of a gas tap, then what gets produced is generally what is chemically most stable. Ingredient A will combine with Chemical B to produce C and a bit of D, if the combination of C and D is more stable than what went before. In astrochemistry, however, time is in great supply, whilst energy – certainly in the cold gas clouds – is not. For example, the combination of the three atoms H, C and N can exist in two forms: Carbon and Nitrogen are always joined strongly together; the Hydrogen atom can be joined to either C or N. So H-CN or H-NC are both possible. To get the Hydrogen to swap ends requires overcoming an energy barrier. In the chemical laboratory, reactions to make Hydrogen Cyanide, HCN, inevitably end up as has just been written – HCN with the Hydrogen atom attached to the Carbon atom, which, in turn, is attached to the Nitrogen atom. HCN is the stable form, and any of the unstable form HNC that might be formed is quickly turned back into HCN because there is enough energy to overcome the energy barrier to do so.

Not so in the ISM. At one point during the course of the chemical reactions that make Hydrogen Cyanide or Hydrogen iso-Cyanide, as the less stable variant is called, there is a Hydrogen atom attached to both the Carbon and the Nitrogen atom. If the Hydrogen attached to Nitrogen breaks off first, so be it: normal HCN is formed. Should it happen that the Hydrogen attached to Carbon breaks off first, then HNC – iso-Cyanide – is the result, and there is not enough energy to overcome the reaction barrier to get the Hydrogen to switch ends. Both HCN and HNC are well known inhabitant of the ISM in our galaxy and others, however over long periods of time, HNC should eventually reform into its more stable HCN form. Left to its own devices, the HNC/HCN ratio is a ticking clock, counting off the millennia.

The fact that energy is at a premium shows up in other ways, too. One such is the "fractionation effect". This refers to the fraction of molecules that have heavy Hydrogen – Deuterium, $_1^2D$ – in their chemical makeup. By rights, since there are only two Deuterium atoms for every 100,000 normal $_1^1H$ Hydrogen atoms, chemicals that include Hydrogen should only have two Deuterium-containing molecules for every 100,000 "normal" ones. Quantum mechanics, so important for understanding how molecules form and stay formed in the first place, has a trick up its sleeve, however; and this trick is called Zero Point Energy.

Quantum Mechanics says that a molecule cannot sit at the very bottom of its electronic potential energy valley. This follows from a central quantum tenet called the Heisenberg Uncertainty Principle, named for the German physicist Werner Heisenberg. Heisenberg found that the equations of Quantum Mechanics gave a different answer if you tried an operation, a measurement, to find the position of a particle followed by an operation to find its momentum than if you carried out the operations in the reverse order: momentum first, followed by position. This always gave rise to an uncertainty in the position multiplied by an uncertainty in the momentum that could never be less than Planck's quantum constant h, divided by two times pi. Matter at the quantum level always existed in a state of uncertain restlessness.

Paul Dirac, who first introduced the concept of anti-matter into Quantum Mechanics, explained this as showing that the state of no motion is the state of no particle at all. This was a strange and shocking conclusion, since it went against everything that Classical Mechanics had taught scientists over the past two centuries or so. They had been used to visualizing the path of a cannon ball as it flew from the mouth to its target as a series of instantaneous "steps" for which both the momentum (given by the mass of the ball multiplied by the speed at which it was traveling) and position were exactly known. After all, you had a good idea of the cannon ball at rest on the deck of a ship or on the shore battery, and this could be transferred to visualize the ball in flight after it had been fired. But Quantum Mechanics would not allow that; if you knew *precisely* the position of a particle, its momentum was utterly undetermined and undeterminable, and *vice versa*. Matter and motion were therefore simply inseparable – matter-in-motion.

Roll a boulder down a valley and, unless it gets stuck on a ledge, it will end up at the bottom, resting there peacefully until someone tries to roll it up the hill again. We might expect that a molecule will end up resting at the bottom of our potential energy surface valley, with its atoms at rest with respect to one another. Unfortunately, this would violate the Uncertainty Principle: you could sit alongside Hydrogen Atom 1 of the diatomic Hydrogen molecule H_2, for example, and you would know exactly where Hydrogen Atom 2 was to be found and what its momentum would be – in this case, zero. The uncertainty in position would be zero, and the uncertainty in momentum would be zero, and zero times zero is always zero, a condition not allowed by Heisenberg's Principle.

Instead, the Uncertainty Principle leads to the conclusion that even in its lowest vibrational energy level – its Ground State – at a temperature of absolute zero, the molecule must be in a state that has half a quantum of energy associated with it. The molecule is not right at the very bottom of its energy valley, but a little way up it; it is as if the valley has a burbling energy stream at its bottom, and, to keep its feet from getting wet, our molecule has to have a vibrational bridge even at its lowest energy level. This residual energy is the Zero Point Energy.

For molecules that are chemically similar, like the diatomic Hydrogen molecule H_2 and its Deuterium-substituted counterpart, HD, the energy of the lowest energy level bridge turns out to be lower down the valley for the heavier molecule. So HD is favored when the temperature is low and energy is in short supply. This is true also for our guide: Deuterium-substituted H_2D^+ is favored over normal H_3^+ at low temperatures. It also turns out that it is always energetically preferable to hand on a Deuterium ion D^+ than a normal H^+ ion when it comes to making more complex molecules like *heavy* Water HDO, rather than H_2O, or even *heavy* Hydrogen Cyanide DCN, not HCN. In the coldest clouds, such as the Taurus Molecular Cloud, this enrichment process can mean that as many as 10% of all molecules contain Deuterium instead of Hydrogen, 5,000 times the normal abundance of Deuterium.

Ion reactions involving our guide H_3^+ are important in what is called the *diffuse* interstellar medium, where gas densities are probably not more than a few or a few tens of millions of atoms and molecules in every cubic meter. Our guide is also active in the outer layers of dense gas clouds, where the higher numbers of particles – perhaps a thousand times more than in the diffuse ISM – absorb the ionizing radiation more rapidly. At the very centers of the largest dense molecular clouds, however, where the temperature may be as low as 10K, it is hard for the gas to become ionized to any great degree. There, other agents are needed.

Not all of the reactions occurring in the interstellar gas clouds have to go through the hands of our guide. To ionize atomic Hydrogen takes 13.6 electron volts: an electron volt, eV, is the energy that a single electron obtains when it passes through a voltage of just one volt. (Nearly 3½ billion, billion electrons per second have to be passed back and forth though the American 110 V mains system to produce as much energy as an average 60 W household light bulb.) It takes slightly more energy, 15.4 eV, to ionize molecular Hydrogen to form H_2^+ and an electron. Since atomic and/or molecular Hydrogen is pretty much everywhere, that means that ionizing radiation – ultraviolet photons or cosmic rays – with

13.6 eV or more will be absorbed in ionizing Hydrogen. Helium, the next most abundant element in the Chemical Cosmos needs almost twice as much energy as Hydrogen to ionize it, so it loses out when energy is scarce. Next in the list of cosmically abundant elements is Oxygen, which also requires about 13.6 eV to ionize it, though just a little more than atomic Hydrogen. So by the time H has been ionized there is no radiation around to produce the O^+ cation. Since the energy required to produce H^+ and O^+ are virtually the same, though, an Oxygen atom can lose an electron to a Hydrogen ion, producing the highly reactive Oxygen ion. This starts its own chemical pathways.

Next in line in cosmic abundance is Carbon.

In the constellation of Leo, the Lion, is the oddly named star IRC +10216. At a wavelength of 10 μm, it is the brightest object in the sky. IRC +10216 is a mature star, already on the way to becoming a white dwarf, throwing off the outer layers of its own atmosphere. This stellar envelope is formed of a dense chemical mix, "sooty" as a sputtering candle flame and rich with Carbon. More interstellar Carbon-based molecules have been found around IRC +10216 than anywhere else. Carbon chemistry may benefit from the presence of our guide, H_3^+, but is not dependent on it, for Carbon requires only 11.3 eV to ionize it. Thus any radiation with an energy below 13.6 eV but above 11.3 eV can ionize Carbon; Carbon chemistry can occur deeper into dense molecular clouds than our guide can take us.

Carbon is a remarkable atom, the basis of organic chemistry. Its compounds are the most abundant in the ISM so it indeed deserves a little diversion, since its chemistry is so important for us all.

Sitting right in the middle of the first of the chemical "octaves", Carbon has four valence hands with which to catch onto other atoms, and it is both generous and greedy with those hands. In the simple molecule Carbon Dioxide, the greenhouse gas, Carbon uses two hands to hold onto each of the two Oxygen atoms, which are two-handed guys. Two bonds are formed between the Carbon atom in the middle and Oxygens on the outside, each Oxygen atom making use of two of its electrons to share with the central Carbon atom, which returns the complement on both sides. Hands go into hands and everyone is satisfied.

In poisonous Carbon *Monoxide*, however, Carbon goes hand-in-hand with each of the two hands of the Oxygen atom, sharing two of its electrons. Then there is a problem: the Oxygen atom has its full octet of electrons, thanks very much, but Carbon has only six. To solve this, the Carbon atom insists on a share of two more of the Oxygen octet, without making any return. Now each atom has its octet and its hands are full; but some hands are fuller than others.

Carbon has a remarkable ability to form molecules in which it combines with itself, over and over again, in a multiplicity of ways. The simplest combination of Carbon and Hydrogen, provided there is enough Hydrogen available – and there usually is – is Methane. In Methane, a single atom of Carbon uses its four valence hands each to combine with an atom of single-handed Hydrogen to form CH_4. Carbon in this form can then daisy-chain: two four-handed Carbon atoms each give up one of their Hydrogen atoms and link together to form Ethane. Ethane's chemical formula is C_2H_6, denoting two Carbon atoms combined with six Hydrogen atoms. But it could be thought of two Methane, or Methyl, radicals, CH_3, joined back to back as H_3C-CH_3. Next in the family album is Propane. In Propane, one of the CH_3 groups has given up a Hydrogen atom and combined with another Methyl radical giving it an overall chemical formula of C_3H_8, denoting three Carbon atoms combined with eight Hydrogens. This could also be written $H_3C-CH_2-CH_3$. There are now three Carbon atoms joined to each other by a single hand each, and making use of their free hands to join with single-handed Hydrogens.

Take a Hydrogen away from one of the end Carbons, add another Methyl radical, and you get Butane, C_4H_{10}, four Carbons in a chain and combined with ten Hydrogens, or $H_3C-H_2C-CH_2-CH_3$. This process can be repeated over and over to build Pentane, Hexane, Heptane, Octane, etc. The names of all of these compounds end in -ane, and together they form a series of compounds that are called Alkanes. If there are n Carbon atoms in an Alkane, there are twice as many Hydrogen atoms plus two – $2n+2$. Compounds containing just Carbon and Hydrogen all fall into the grand family of HydroCarbons. The Alkanes are termed "saturated" HydroCarbons because each atom is joined to its neighbors by just one of its valence hands, forming a single chemical bond.

If, however, a Hydrogen molecule is removed from an Alkane, an Alkene is formed. So Ethane, C_2H_6 or H_3C-CH_3, becomes Ethylene or Ethene, as it is now known, C_2H_4 or $H_2C=CH_2$. The = shows that the Carbon backbone is now joined with two bonds and Alkenes all have two Carbons joined by a double bond. Propene has the formula C_3H_6 and can be written $H_3C-CH=CH_2$. Butene has the formula C_4H_8, and so on. Alkenes have exactly twice as many Hydrogen atoms as there are Carbons in the chain. Take another Hydrogen molecule from Ethene, C_2H_4, and Acetylene, or Ethyne as it has been renamed, is formed – C_2H_2 or $HC\equiv CH$, where the Carbons are now joined by three bonds – a triple bond. Alkynes such as Ethyne, Propyne, Butyne and so on all have twice as many Hydrogens as Carbons, minus two for the Hydrogen molecule that has been taken away from the corresponding Alkene. Alkenes and Alkynes are all called unsaturated HydroCarbons because their Carbon atoms could give up their double or triple bond status and combine with more Hydrogen atoms should they become available.

All elements, even the rather unreactive, so-called noble gases like Helium, Neon, Argon, and Xenon, do chemistry of some sort or another; but for Carbon they could have written the James Bond theme "Nobody does it better". Carbon is incestuous – combining with itself – and promiscuous – combining with pretty much any element that comes its way with equal enthusiasm. Since Oxygen is more abundant than Carbon in the universe, and Nitrogen is next after Carbon, chemicals that combine Carbon, C, Hydrogen, H, Oxygen, O, and Nitrogen, N, the CHON compounds, are the most prolific in the Chemical Cosmos. Put Oxygen into a HydroCarbon and Alcohols, Aldehydes, Ketones, and Acids are formed, among others; put Nitrogen in and Cyanides and Amines result; put Oxygen and Nitrogen in to a HydroCarbon and amongst the compounds that result are Amino Acids, the building blocks of proteins.

The complicated nature of Carbon chemistry also arises in a property known as *isomerism*. The overall chemical formula of two or more compounds may be the same, but because they have shape and structure they are actually different: they are *isomers*. With the simple Alkanes, this happens as soon as you reach Butane in the series – C_4H_{10} – four Carbon atoms and ten Hydrogen atoms.

This comes in two different structures. Normal Butane has the four Carbons all in a line: $H_3C-H_2C-CH_2-CH_3$. But now let us suppose the Carbons are numbered 1 through 4 from left to right in the chain; we could create a branch on one of the two middle Carbons – say Carbon-2 - by removing a Hydrogen atom and giving it directly to the other middle Carbon, Number 3. That would mean that Carbon-3 had no valence hand left to link to Carbon-4, but Carbon-2 *would* have a free hand to link to the now homeless Carbon-4. So Carbon-2 is now linked directly to Carbon-1, Carbon-3 and Carbon-4; its fourth valence hand links to a Hydrogen atom. This structure can be written $HC(CH_3)_3$ indicating that one Carbon is attached to three Methyl radicals plus one of its original Hydrogens. It still has the overall chemical formula C_4H_{10} – four Carbons and ten Hydrogens – and it has no double bonds, so it is still a fully saturated Alkane. By definition it must be Butane, but not as we used to know it. It used to be called *iso*-Butane because it was a branching isomer of normal single-chained Butane. Today it is called 2-Methyl-Propane to indicate that it is a Propane molecule with an additional Methyl radical attached to Carbon-2 in the chain.

Manipulating the Carbons and Hydrogens in an Alkane like Butane does little to the chemical properties; with more Carbon atoms in the chain it is possible to create more and more different structures, with little impact on the chemistry. Take, however, the compound made up of two Carbon atoms, four Hydrogen atoms and an Oxygen atom, C_2H_4O. This could be Acetaldehyde, from the same family as Formaldehyde and can be written H_3C-CHO, which is a Methyl radical attached to an Aldehyde – CHO – group. In the Aldehyde group the Carbon atom forms two bonds to Oxygen, which is a two-handed atom, and one with the Hydrogen atom, which gives it one to link to the Methyl radical.

We could pass a Hydrogen atom right down the chain from the Methyl radical to the Oxygen atom, however. Oxygen, being two-handed, has one tied up with its link to Hydrogen, so only one is available for the Carbon atom next to it. That leaves that Carbon short-handed but because the Carbon of the Methyl radical has given up one Hydrogen, it, too, is short-handed. The two Carbons happily link up to form a double bond. The overall chemical formula is unchanged – still C_2H_4O – but the structure has gone from

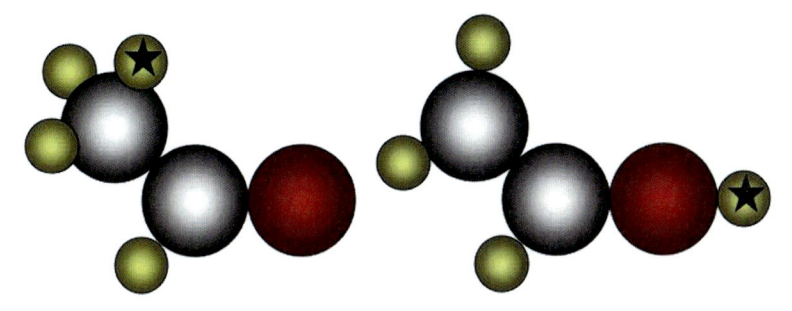

FIGURE 4.5 Carbon forms isomers, molecules with the same overall chemical composition, but with different arrangements that have different properties. On the left is Acetaldehyde. The starred Hydrogen atom in the top left moves onto the Oxygen in the molecule on the right to form Vinyl Alcohol. (Key: Carbon atom – black; Oxygen atom – red; Hydrogen atom – yellow.): *credit – Steve Miller.*

H_3C-CHO to $H_2C=CH-OH$. This is a very different compound from Acetaldehyde called Vinyl Alcohol. Alcohols combine with Acids to form salt-like compounds called Esters; Aldehydes do not. *Plus ça change plus cela n'est pas la même chose* (The more things change, the more they *do not* stay the same, in this case) (Figure 4.5).

Both of these forms, however, exhibit another property existing in many molecules but which, once more, Carbon compounds illustrate in the greatest variety – polarity. Oxygen as an atom, for example, has a stronger pull on the electrons that form chemical bonds than the Carbon atom. In Vinyl Alcohol, that means the Oxygen atom carries a slight negative charge and the Carbon next to it is slightly positive. The double bond between the two Carbon atoms in Vinyl Alcohol is also a very reactive site for chemistry to occur; therefore atoms and molecules that have electrons to spare – anions or molecules like Ammonia, NH_3 – can form attachments to Vinyl Alcohol. Vinyl compounds can also combine with each other, in a process known as *polymerization*. If instead of Vinyl Alcohol we have Vinyl Chloride, the well-known plastic PolyVinyl Chloride, or simply PVC, results from polymerization. Aldehydes are also polar molecules since the Oxygen atom there also becomes slightly negatively charged. When the simplest of this family, Formaldehyde, polymerizes, Sugars result.

Overall, polar molecules are more reactive than non-polar molecules. That means that while their chemistry is not as active as that of charged molecules, ions like our guide H_3^+, it can still occur where the energy needed to get over the reaction barrier is in short supply.

Chemical compounds, unlike the individual atoms from which they are formed, have shape: they create three-dimensional structures. Once more, nobody does it better than Carbon. If Carbon forms a triple, three-handed, bond with another atom – with another Carbon, as in the Alkynes, or Nitrogen, as in the Cyanides – its final valence hand can only stick out in the opposite direction. The result is that part of the molecule has to be linear. Ethyne is straight, and Hydrogen Cyanide is also linear. Quantum mechanically, Carbon is said to be in an *sp* configuration. When Carbon forms just one double bond and two singles, the three bonds pretty much point out to the corners of an equilateral triangle, with an angle of about 120° between them. Quantum mechanically, Carbon is said to be in an *sp²* configuration. The three atoms surrounding an *sp²* Carbon atom all lie in the same plane, so Ethene is a completely flat molecule. When Carbon has four single bonds, these all point to the corners of a tetrahedron with an angle of 109½° between them. Such Carbon atoms are said to be in an *sp³* configuration, and they just will not lie flat. *sp³* Carbon atoms always give rise to three-dimensional molecules, so Methane has the "shape" of a tiny tetrahedron.

sp³ Carbon atoms also lead to another property – chirality, or molecular handedness. Lay your arm and hand flat on the table, and splay the thumb away from the rest of your digits. If we said your wrist was an atom, then your arm, thumb and remaining fingers more or less point in the direction of the three bonds of an *sp²* Carbon. Do that with both arms and hands, and then bring them together as if you were praying. Left arm fits on right arm, left thumb on right, and left fingers on right fingers. Now splay the index and middle fingers such that along with the thumb, they all point in different directions (Figure 4.6). (Note that the other two fingers can be tucked into your palm as they are not needed.)

Unless you are double-jointed, you can probably get them all to point at about 90° from each other, not quite the 109½° of the *sp³* tetrahedral Carbon, but close enough to illustrate the point.

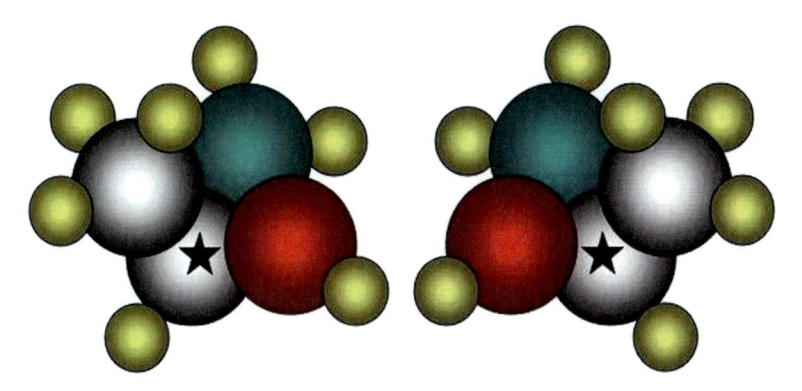

FIGURE 4.6 sp³ Carbon forms *optical* isomers, molecules with the same overall chemical composition, but with different arrangements. The central starred Carbon atom is surrounded by four different groups – a Hydrogen atom, a Methyl group, an OH group (Alcohol), and a NH_2 group (Amine). The left and right-hand molecules are mirror images of one another. (Key: Carbon atom – black; Oxygen atom – red; Nitrogen atom – blue; Hydrogen atom – yellow.): *credit – Steve Miller.*

Do this for both hands, and once more bring the arms together. You can get the first and middle fingers to point in the same direction, but the thumbs either point towards or away from one another. Yoga experts can try doing this behind their backs, in the lotus position, or any other limb-wrenching position they like, but they cannot get left arm, fingers and thumb all to point in the same direction as the right. Yes, the left hand knows what the right is doing, but it just cannot do it. Mimicking an *sp³* Carbon, the left and right hands are mirror images of each other (Figure 4.6).

Chirality does not matter if an *sp³* Carbon is attached to four Hydrogen atoms, as in Methane, because the Hydrogen atoms are indistinguishable and you can put a right-hand Methane by a left-hand Methane and cannot tell the difference. Substitute one of the Hydrogens for another single-handed atom or molecule, and left hand and right hand can still be maneuvered to look alike. Similarly, another substitution makes no difference, but when all four of the Carbon bonds are each linked to different atoms or molecules, then the resulting right-hand and left-hand molecules cannot be positioned to look alike; they, too, are mirror images of one another. This property does not often show up, but left-hand and right-hand molecules interact differently with polarized light.

In living organisms on Earth, there is a preference for left-handed Amino Acids, for reasons that are still not fully understood.

Added to its chain-forming properties is Carbon's ability to form rings. Obviously Methane itself cannot form a ring, as there is only one Carbon atom, nor can Ethane, which has but two and has to be a straight line compound. Any one of the Alkanes from Propane, which has three Carbon atoms in its chain, onwards can be found in a ring formation, however. Note that in forming a ring, two Hydrogen atoms have to be lost so that the Carbons at the end of the chain have a free valence hand to join onto one another. This results in *cyclo*-Alkanes having the general formula of exactly twice as many Hydrogen atoms as Carbons, like the chained Alkenes – another example of the multiple isomerism of Carbon compounds. *Cyclo*-Propane is a triangle with the angles between the chemical bonds at 60°, causing a great strain on the sp^3 Carbons whose bonds like to stick out at 109½° to one another. *Cyclo*-Propane has a tendency to pop open again, given the slightest provocation; *Cyclo*-Butane forms a puckered square, the two Carbons opposite each other sticking "up" and the other two pointing "down". The bonds are closer to 109½° apart and *cyclo*-Butane is a happier molecule than its smaller sibling, but is still rather strained and tense. *Cyclo*-Pentane and *cyclo*-Hexane are quite comfortable, on the other hand, since their bonds can more or less reach the magic 109½°.

The six-Carbon ring of *cyclo*-Hexane shows the true three-dimensionality of Carbon molecules. It has two forms that flip easily into one another but nonetheless make for rather different chemistry. *Cis-cyclo*-Hexane has a shape a bit like a boat. Carbons-1 and -4 in the ring, the ones opposite each other point "up" compared with Carbons-2 and -3 and -5 and -6. *Trans-cyclo*-Hexane, on the other hand, has Carbon-1 pointing "up" and Carbon-4 "down", a bit like a crude deck-chair. The *cis* and *trans* labels come from the Latin words meaning "on this side of" and "on the other side of". In the days of early Rome, *cis*-Alpine tribes were friendly, or at least had had the stuffing knocked out of them; *trans*-Alpine tribes were definitely not to be trusted. Compounds derived from *cis-cyclo*-Hexane with chemically active groups on Carbons-1 and -4 have those active groups close enough to react with one another, or simultaneously with two other active groups on another molecule. On *trans-cyclo*-Hexane, the reactive groups are kept isolated from one another, and will usually only react with active groups on other molecules one at a time. In Carbon chemistry, shape really matters and, quite literally, our lives depend on it.

If sp^2 Carbons are put into a ring the result is what is called *aromatic* compounds since they are almost always accompanied by strong (and not always pleasant) smells. In these, anything from three to several Carbon atoms – sometimes with other elements mixed in – join together; some Amino Acids and what are called the "nucleobases" of RiboNucleic Acid (RNA) and Deoxyribo-Nucleic Acid (DNA) require these rings. The best known of these ring compounds is Benzene, with six Carbon atoms arranged in a neat hexagon. Benzene's six sp^2 Carbons should, officially, have a double-handed link to one Carbon neighbor and a single bond to the other. So Carbon-1 could form a double bond to Carbon-2, forming a single bond to Carbon-3. Carbon-3 is then free to form a double bond to Carbon-4, which attaches itself singly to Carbon-5; Carbon-5 double bonds to Carbon-6, which completes the ring with a single bond to Carbon-1. But it could just as easily go the other way round: 1–2 single; 2–3 double; 3–4 single; 4–5 double; 5–6 single; 6–1 double. These are equivalent, *resonance*, structures and early attempts to explain the bonds in Benzene made use of this idea. Full quantum mechanical calculations, however, have a ring of bonds with a buzz of electrons *delocalized* equally across all six Carbon atoms like a swarm of bees. Delocalized bonding, as our guide H_3^+ knows only too well, is a way of keeping all atoms in a molecule equally served by the electrons that bind them together.

When Hydrogen is very abundant, a single Carbon atom will tend to form Methane; where it is not and where there are rather few molecules – often the case in the Interstellar Medium – then radicals such as the Methyl radical CH_3 can be formed. On its own, the Methyl radical is hungry and reactive; its Carbon atom has a free valance hand waving to grab whatever might pass its way. Carbon atoms will also form radicals with just two, and even one, Hydrogen. Where protons are available, everything can become protonated – Methane becomes CH_5^+ for example, and the CH_3^+ radical ion is well-known. Knock a proton out of Methane and the negative anion CH_3^- can be formed. What works for Carbon also works for Oxygen, Nitrogen and other species. Water – H_2O – gives rise to the radical OH, the cations H_3O^+ and OH^+ and the anion OH^-. Ammonia (NH_3) produces radical NH_2, cation NH_4^+ and anion NH_2^-; however, none of the other elements beats Carbon for complexity.

Now back to the clouds between the stars and our tour down the broad river of cosmic chemistry.

Ionized Carbon, C^+, is a very reactive species; and like our guide, H_3^+, its electric charge makes it particularly attractive to other molecules in cold clouds, where there is no energy to get neutral cooking underway. One of its key reactions is to latch onto molecules of Ammonia or Ammonia-like molecules to form Hydrogen Cyanide. This Hydrogen Cyanide forms an astronomical feedstock for Carbon. Starting with a Cyanide molecule containing just one Carbon atom, further reactions can lead to the addition of Carbon after Carbon into the chain. To date, the longest chain of this type that appears to have been discovered in space has 11 Carbon atoms in a line, with a Hydrogen atom at one and a Nitrogen atom at the other. In between, the Carbon atoms are content to hold hands and share electrons. In the ISM, these Cyanide-containing molecules have another crucial property – they can polymerize, making longer chains that are, in themselves, very reactive. Such Cyanide-bearing chains hold the key to creating Amino Acids, the first steps to building the Proteins that are life-forming molecules.

Carbon chemistry in space is enormously rich; of the 144 molecules detected somewhere in the ISM by 2006, some 70 years after the first molecules were detected in space, 75% had at least one atom of Carbon. Benzene is one of the several Carbon ring-molecules that have been detected in interstellar clouds, or in the envelopes of stars going through their final stages of evolution. Carbon can also form itself into chains or rings of rings, creating what are known as poly-cyclic aromatic hydrocarbons or PAHs for short. Simple PAHs such as Anthracene and Naphacene have three and four rings chained together, respectively, in a straight line. Other PAHs have their rings bent or folded in on themselves. One particularly symmetric formation has six Benzene hexagons linked together to form a hexagon of hexagons, and is known as

Coronene. These PAH molecules, in various guises, have been proposed as the agents causing the visible and ultraviolet Diffuse Interstellar Bands first detected by Mary Lea Heger. Despite PAHs, in general, proving quite a good fit as the family of chemicals causing the DIBs and other ubiquitous features in the spectrum of the interstellar medium, it has proved difficult to point the finger with certainty at any individual culprit.

The Red Rectangle is a nebula of gas close to the southern constellation of Monoceros the Unicorn first discovered in 1973. As its name suggests, it has a rectangular shape, with a cross of bright emission extending from the aging star at the center out to its corners. The ubiquitous DIBs show up clearly in the Red Rectangle; unlike their usual occurrence, however, these bands are not dark lines across a star's spectrum, but are bright lines shining out into space. In 2004, the American Astronomical Society was told that the PAHs Anthracene and Pyrene had shown up in the ultraviolet spectrum of the Red Rectangle. If true, these would be some of the most complex Carbon molecules yet found in space. But for many astronomers, while PAHs remain among the "possibles" or even the "probables" of interstellar chemistry, they have not yet made it into the "definites".

Carbon's ring-forming is not confined to making essentially flat molecules; groups of rings can bend round on each other to form nanoscale "cages". In 1985, Harry Kroto, of Sussex University in southern England, and Rich Smalley, of Rice University in Texas, discovered a new form of Carbon, to add to graphite and diamond. It consisted of 60 Carbon atoms arranged in pentagons and hexagons like a soccer ball. The two named their discovery Fullerene after the architect Richard Buckminster Fuller, who used geodesic domes as the focus of his structures; Kroto and Smalley, along with their Rice colleague Bob Curl, were awarded the Nobel Prize for Chemistry in 1996 for this discovery. Fullerene and its cousins now form the basis for the vibrant research area of "nanotubes", the chemistry of making longer and longer versions of the molecule so that it no longer resembles a soccer ball, or even an American football, but a tiny tunnel of Carbon atoms that may cage metals and other chemical species in the hope of producing a generation of electronics at the genuine nanoscale. Fullerenes have reportedly been detected in the ISM.

As things get bigger, new properties often arise. David Williams, former president of the UK's Royal Astronomical Society, has waged a long campaign for cosmo-chemists to recognize the importance of "dust" for the Chemical Cosmos. In space, Carbon can also form much extended clusters of atoms coated with Hydrogen, called Hydrogenated Amorphous Carbon – HAC to its friends. The ability to build extended, three-dimensional structures gives Carbon the ability to team up with other molecules that can easily solidify such compounds of Aluminum and Silicon to form grains of interstellar dust.

Most probably, grains do not form in the ambient ISM but in special environments, such as in the outflows of chemically rich gas given off by aging stars like that at the center of the Red Rectangle or the Carbon-star IRC +10216 that we met earlier. In these outflows, gas densities are high in comparison with the average with some 10 billion billion molecules in a typical cubic meter, "only" a million times less dense than the air we breathe. For a star rich in Carbon, the grains that form will themselves be Carbon-rich. In most stars, however, Oxygen is significantly more abundant than Carbon, so what Carbon there is will be locked up as Carbon Monoxide, or CO. Then the grains will be rich in Aluminum and Silicon.

Chemistry can take place on the surface of grains even in the absence of radiation, which is normally needed to get chemical reactions to start in the gas clouds of the ISM. Once formed, grains act as minute chemistry laboratory workbenches, encouraging individual atoms and molecules to settle down and see what they can get up to. Hydrogen atoms can get together to form diatomic Hydrogen molecules without having to go through the palaver of getting ionized and giving off photons because the grain surface brings them into a convivial intimacy, encouraging bonding and acting as a sink down which they can funnel any of the excess energy that might tend to push them apart again. In its own leisurely way, the grain will then radiate that energy back out into space, not as a bond-busting ultraviolet photon but as a warming series of infrared- and micro-waves. With the temperature around 70K, more than 200K below the freezing point of Water, CO, ubiquitous even in Carbon-rich environments, can freeze out onto the surface of grains. From this vantage point, it can start chemical chain reactions that lead to intoxicating mixtures of, amongst other things, Alcohols, Aldehydes and Amino Acids.

None of this happens without resorting to another unexpected result of Quantum Mechanics, however. The grains that start their lives in relative warmth close to the dying star that gave rise to them do not stay there. Over hundreds of thousands and millions of years, they may travel tens of millions of kilometers into the coldest reaches of the galaxy. Losing heat by giving off infrared photons, they gradually settle down to the temperature of the gas that surrounds them. When an atom of Hydrogen lands on the surface of the grain, it too cools down so that it has very little energy to move. So unless it just happens to land right next to another Hydrogen atom, how will it team up and form a molecule? And even when a simple molecule forms, how will that team up with other atoms to form complex molecules?

Left to Classical Mechanics, individual atoms would simply stay where they landed on the surface of a cold grain and there would be no chemistry. But atoms and molecules, in the quantum picture, are not just particles but also waves; their wave functions are not tied to just one location in space but are delocalized. As such, there is a better than zero probability that the atom or molecule could be found anywhere on the grain's surface, a quantum slime sliding from place to place, seeking out similarly slimy partners with which to form a molecular alliance. In this way, atoms and molecules are able to overcome the energy barrier that prevents them moving across the grain. This "tunneling", as it is known, is a fundamental feature of Quantum Mechanics, an essential property of wave-particles like electrons, atoms and molecules, found everywhere in the universe. So powerful is this ability to tunnel through seemingly insurmountable energy barriers that particles can even escape the extreme gravitational clutches of Black Holes, those cosmic sinks down which anything that ventures too close is bound to fall.

Despite their generally low temperatures, grains contribute significantly to the Chemical Cosmos. Indeed, many of the larger organic molecules owe their existence to dust and the surface it provides for complex chemical reactions. These organics may even be "chips off the old block", bits blasted off grains by collisions or powerful radiation.

As well as forming ideal workbenches for interstellar chemistry, cooking even without the aid of starlight or cosmic rays, grains gradually become covered with ices. Water ice, of course, forms as soon as they reach a temperature of 273.15K above absolute zero (the zero of the Centigrade scale). At 57K below that, Carbon Dioxide – CO_2 "Dry Ice" – can freeze out, followed by Ammonia 20K lower. Gradually the grains build up coat after coat of chemical ice, and gradually the surrounding gas loses its chemical richness. It is out of this mix of gas and ice-encrusted grains that the next generation of stars and, significantly for us, planets form.

Key to our understanding of these processes has been the information gathered over the decades from the spectra of molecules, from radio waves, microwaves and, especially, from the infrared.

5. Fishing for Molecules

The largest mountain on Earth is the dormant volcano of Maunakea, in Hawaii, which reaches up 4,200 m above sea level. Although it is not the world's highest point – that honor still goes to Mount Everest in the Himalayas – from its base on the Pacific Ocean to its tip, Maunakea measures roughly 10,200 m. In tourist brochures, the island state of Hawaii is usually portrayed as a land of beaches and palm trees, with surfers skimming giant waves and dancers swaying in the tropical breeze. (It is the home of the aloha shirt and the grass skirt, for example). Whales, dolphins and sharks patrol the waters off the islands. It is the place to go to get some winter sunshine when the snows and rains have set in or to bake during the summer months.

Hawaii is also one of the most naturally diverse and active places in the world. The mountain range of which Hawaii is part stretches for over 2,500 km, from Kure and Midway Island, halfway between Japan and the USA. Most of the range is now submerged but at least some of the tiny islands in the northwest may – millions of years ago – have been as high above sea level as Maunakea is now. The island chain that makes up the state is the result the Pacific plate of Earth's crust drifting from southeast to northwest over a hotspot, a region where vast reservoirs of molten magma are to be found. As a result, a series of volcanoes have been formed, been active, become extinct, and eventually sunk back into the Earth's crust, while all the time being worn away by wind, rain and ocean for periods of some tens of millions of years.

The oldest island in the Hawaiian chain, Necker, which covers a mere quarter of a square kilometer, is around 10 million years old, according to geological dating techniques which make use of the decay of a radioactive Potassium isotope into Argon. The youngest of the islands, Hawaii itself, is only 600,000 years old. Hawaii is also the largest island, more than 10,000 km^2 in area and getting larger every day as its still active volcano, Kilauea, pours molten lava into the sea at the south of the island.

Much smaller (by a factor of several) than Hawaii itself are the islands of Kauai, Oahu and Maui. Oahu is the most densely populated, with the state capital Honolulu. Nihoa, Niihau, Lanai, and Kahoolawe are the other islands of significant size; still beneath the surface of the ocean and growing steadily is Loihi, mapping the continued drift northwestwards of the Pacific plate, and the southeastern extension of the chain.

All of the main islands are sub-tropical; their latitudes range from 22° North (Kauai) to 19° North (the south tip of Hawaii). Stuck in the middle of the Pacific Ocean between longitudes West 155° and 160°, they make up the most geographically isolated landmasses on Earth. Ocean waves pound the shore, creating beaches of salt-and-pepper, white coral and black lava sand sprinkled with tiny crystals of shining green olivine. Sea-going turtles haul out on the beaches to rest, getting rid of parasites and to lay their eggs.

Mostly the weather comes into Hawaii from the east, born on the prevailing Trade Winds. The Trades are warm and wet as a result of their uninterrupted passage across a thousand miles of sea. As they hit the islands, particularly the islands with high mountains like Maui and Hawaii; they dump their moisture as rain varying from light mists to torrential downpours. The results are spectacular.

It depends how you count them, but naturalists reckon that the state of Hawaii has more than 80% of the world's recognizable biospheres. These range from dense tropical rainforest, on the eastern, or windward, slopes of the islands to near-desert on the leeward sides. The enormous mountains of Maunakea and Mauna Loa (the white and long mountains, respectively) on Hawaii and Haleakala (the House of the Sun) on Maui also give the islands the kind of climates known to Europeans living in the Alps. Then there is nearly everything in between...

Hawaii has probably long been associated with astronomy. It is the northernmost part of what is known as the Polynesian Triangle. This culturally connected region is centered on the Society and Marquesas Islands and stretches from Rapanui (Easter Island) in the southeast to Autearoa (New Zealand) in the southwest. Sometime in the fourth century AD, at the time the Roman Empire was at its zenith, or maybe the fifth century AD, when it fell, ushering the Dark Ages into Europe, brave voyagers set out north over the Pacific Ocean from the Marquesas. They were guided on their voyage to Hawaii by winds and currents, and their close observations of bird flight and marine life.

There is evidence that the first humans to arrive at Hawaii were also navigating by their knowledge of the stars. When Europeans first came into contact with the Hawaiians in 1778, with the arrival of Captain James Cook and his ships, the Resolution and the Discovery, they found that there were local names for stars and constellations; Arcturus in Bootes was Hoku Lea, the happy star, perhaps because it had been used as a navigational aid on the "original" voyage to Hawaii. Scorpio, or part of it, was Maui's Fishhook, and the legend was that the demigod Maui had used it to pull the islands up from the bottom of the ocean.

Although they came with a heavenly purpose, astronomy did not seem to be a preoccupation of the missionaries who "westernized" the islands from the 1820s onwards. It was nearly a century after the arrival of Cook that the first recorded astronomical observation was made in Hawaii. According to Walter Steiger, late Professor of Astronomy at the University of Hawaii, this was the observation of the transit of Venus across the face of the Sun, on December 8, 1874. Ten years later, at the instigation of King Kalakaua, an astronomical telescope was installed at Punahou School on Oahu, the first record of one in Hawaii.

There was great public interest in the 1910 appearance of Halley's Comet; Hawaii was ideally placed for spectacular views of it. The Kaimuki Observatory was set up on a site near Honolulu's iconic Diamond Head for the purpose of observing the comet, and the Astronomical and Astrophysical Society of America organized a special expedition to Hawaii. By the end of World War 2, the interest in astronomy had grown significantly on the islands, ushering in the formation of the Hawaiian Astronomical Society. Steiger himself was instrumental in bringing major astronomical studies to the state. Impressed by the site of Haleakala on the island of Maui, which was already in use for radio astronomy, he proposed building a solar observatory there. The project was to take over a decade of dedicated lobbying and fundraising, but Steiger eventually saw his observatory open in January of 1964.

Astronomers like to place their telescopes high in Earth's atmosphere, if they can. For a start, the atmosphere causes what they call "seeing" which makes stars twinkle and smears the light that comes from them, making images less sharp; the less atmosphere you have to look through, the better. Secondly, depending on where you are, mountains may offer you a chance to get above the clouds that will certainly put a damper on a night's observing. So, for example, the Mount Wilson observatory in southern California, founded in 1904, is at an altitude of nearly 1,750 m. The nearby Mount Palomar observatory is just 30 m lower.

Shortly after it opened, Steiger's solar observatory was visited by the distinguished astronomer Gerard Kuiper, director of the Lunar and Planetary Institute at Tucson, Arizona. Kuiper had heard great things about the conditions on Haleakala and wanted to see for himself whether it could be a suitable site for a new optical observatory. At just over 3,000 m in altitude, Steiger's observatory was usually above the "inversion layer", the point in Earth's atmosphere where clouds form. But not always, and it was sufficiently close to clouds to be affected frequently by fogs and mist.

Looking across the cloud tops, Kuiper noticed that there was a mountain to the southeast that looked even more promising than Haleakala – Maunakea on Hawaii. The summit of Maunakea was clearly well above the cloud tops, and that, he reasoned, would be the place to make a major investment and build a large telescope that would do justice to modern astronomy. The first steps in making Maunakea the world's premier astronomical location were underway, but between dreams and making them reality is a long step.

Surveys of both the Haleakala and Maunakea summits showed that the latter was, as Kuiper had supposed, the better site. At that time, the summit of Maunakea was a pretty inaccessible place, and it took the full weight of the state governor, John Burns, to have a dirt road constructed that would allow the heavy equipment needed to create a modern observatory to be taken to the top. Then there was arranging for funding so the project could go ahead. (Although much of the road is now paved, cars still come off at its steep corners with alarming regularity and with often tragic results.)

Kuiper had persuaded the American space agency, NASA, to put up money to build a 2-m class telescope. NASA then decided

FIGURE 5.1 The summit of Maunakea with the Maunakea Observatory: *credit – Richard Wainscoat, Institute for Astronomy, Hawaii.*

that Kuiper would not have a free run at the money and that others should be allowed to compete for it. In 1959, Hawaii had voted to become the 50th state of the USA meaning that its own university was eligible to bid for funding from NASA, and astronomers there were encouraged to "go for it". So it was local boy John Jeffries of the University of Hawaii who won out over mainland rivals, including Kuiper himself. Jeffries plan was for a telescope that would have a main mirror of 2.2 m (88") in diameter. That made it smaller than the big Californian telescopes, which were of the 3-m class, but the advantages of the Maunakea site were calculated to outweigh the smaller area for collecting light from the stars and planets.

The summit of Maunakea itself was an enormous challenge (Figure 5.1). At 4,200 m, altitude sickness due to lack of oxygen often afflicted workers; an oxygen enrichments system was designed to combat this but was never used. The summit is also bitterly cold in winter, with daytime temperatures not rising above freezing. Snow is a regular winter hazard, along with fog and driving rain. Indeed, despite its subtropical location, during the last Ice Age the summit of Maunakea was covered with glaciers. It must have been a spectacular place; vast glaciers melted by enormous volcanic eruptions to form a world of fire and ice and torrents of water carving deep gullies as they flowed down the mountainside.

Nonetheless, the building of the University of Hawaii's 88" Telescope took just 2½ years, and it was officially opened in June 1970. Jeffries was not entirely satisfied, though, and he felt that "with a little better luck with the weather and a little more prudence on the contractor's part, we could have shaved a year off this time". Big-time astronomy had come to Maunakea, notwithstanding, and it did not take long for the world to notice. By the early 1970s, astronomers in Canada, France, Great Britain, and NASA itself were negotiating to build their telescopes on the summit. Additionally, there was a feature of the high altitude and the dryness associated with Maunakea that was especially attractive – it was ideal for *infrared* astronomy.

As well as the optical universe that we can see with our own eyes, either unaided or through telescopes, there is an infrared universe that for certain objects is much brighter. (Indeed, there are "universes" at all wavelengths of the electromagnetic spectrum – from gamma rays and X-rays through to the longest radio waves.) Nor does infrared radiation simply get through where optical light is stopped. The visible light universe is mainly the universe of atoms; the infrared universe is the universe of molecules, probing matter in a different chemical state. Unlike visible light, however, much infrared radiation is blocked by the Earth's atmosphere. Water may be vital for life on Earth but it is the bane of infrared astronomy, for water vapor is a very good greenhouse gas and absorbs a lot of the infrared wavelength spectrum. So the shorter the column of atmosphere above a telescope and the drier the atmosphere, the greater proportion of infrared radiation astronomers can observe.

The telescopes that were built during the mid-to-late 1970s all set out to exploit Maunakea's unique properties as a site for infrared astronomy. In 1979, 200 years after Captain Cook's death on the shore of Kealakekua Bay, a trio of 3 to 4-m class telescopes all opened up. NASA built its dedicated Infrared Telescope Facility

(IRTF) on a summit ledge all on its own. Britain sited its United Kingdom Infrared Telescope (UKIRT) close to the University of Hawaii's existing telescope, and first in the field, the alliance of Canada (C), France (F), and the University of Hawaii (H) opened their telescope, the CFHT, to great acclaim. These new large, infrared-optimized telescopes were to revolutionize our understanding of the Chemical Cosmos, mapping the heavens in the light of Hydrogen, Carbon Monoxide, and detecting molecule after molecule in ever more exotic locations.

Many of the astronomers who pioneered infrared astronomy on Maunakea made great and original contributions to the field. The infrared universe revealed a world of chemical factories, star nurseries and complex planetary atmospheres. Some of these astronomers, like former IRTF Director Eric Becklin and NASA advisor to the Maunakea Observatory Gerry Neugebauer, have infrared objects named for them. (Becklin also has a road feature named for him; Becklin's Leap is a particularly nasty turn followed by a steep dip in the road on the way to the Institute for Astronomy in Manoa Valley, Honolulu. Take it at speed, as he apparently did, and you will certainly know what Becklin's Leap is all about!) These pioneers also helped put local businesses in Hawaii's main town, Hilo, on the international map. After a night struggling to get recalcitrant instruments to work on the new telescopes, or buoyed by a new discovery, frustrated and/or jubilant astronomers would call in at Ken's House of Pancakes, located opposite the picturesque Banyan Drive, then the only show in town in the wee small hours.

In the three decades that have followed the building of the trio of large infrared telescopes, more diverse and larger have followed. Britain set up its James Clark Maxwell Telescope alongside the California Sub-millimeter Observatory just below the main summit in what has become known as "Millimeter Valley". Caltech's ambitious Keck twin 10-m telescopes followed in the early 1990s. At the turn of the new millennium, two 8-m telescopes also took root. Japan brought an almost industrial style of astronomy with its Subaru Telescope. A consortium of USA, Britain and Canada, together with Argentina, Australia, Brazil, and Chile, opened their Gemini North telescope in 2001; its name reflecting that it had a twin located at Cerro Pachon in the Andes Mountains in Chile.

Pride in their traditions and their islands has never left the native Hawaiian community, despite attempts by some of the missionaries and other haoles to suppress them. This pride underwent a particular revival at the same time as MKO was being developed into a world-leading observatory. There is certainly much pride and interest among native Hawaiians at the development of MKO. Local people flock to the Imiloa Astronomy Center in Hilo for the exhibitions and planetarium shows. They swarm up the mountain to take part in observing at the Onizuka Center, located at 3,000 m and named for the Hawaiian astronaut Ellison Onizuka, who died in the 1986 Challenger Shuttle disaster. It would be wrong, however, to think that the development of the Maunakea Observatory (MKO) and its battery of world-leading telescopes has been universally welcomed and uncontroversial. Although astronomy has brought economic benefit and credit to the state of Hawaii, particularly to the island of Hawaii itself where jobs are hard to find, it is not without its critics.

Maunakea is a sacred mountain to native Hawaiians and many have been dismayed at the rough treatment the summit received at the hands of bulldozers and heavy construction equipment. Moreover the ecosystem of Maunakea is unique; there is even an insect called the wekiu (summit) bug that generates its own antifreeze to withstand the mountain's subzero temperatures. There is obvious concern about the environmental damage generated by astronomical development, and these cultural and ecological concerns have coalesced over the past decade to limit the development of MKO.

The Keck Observatory had originally planned to build six smaller "outrigger" telescopes, essential for its ambition to use the two 10-m main dishes in a coordinated way that would have given them the ability to see astronomical objects only visible to a 100-m telescope. This remarkable feat would have been achieved via a technique known as interferometry, using the difference in the wave front of the light reaching the two main dishes to distinguish very small objects such as planets orbiting distant stars. However, to do this required detailed knowledge of how that wave front varied in the 80 m separating the two main telescopes and that is what the outriggers were designed to monitor. The Hawaii State Board of Land and Natural Resources agreed that Keck could build its outriggers, and work was due to start shortly thereafter.

In November 2004, however, Maunakea Anaina Hou and the Royal Order of Kamehameha I joined forces with environmental activists in the Sierra Club to file an appeal against the board's decision. Two years after the appeal was filed, a judge ruled that it was indeed wrong to give permission for the outriggers in the absence of an MKO "comprehensive management plan" for the summit. The outrigger project died, though, and there is still some debate as to whether or not Maunakea should host the next generation of giant telescopes, which will be 30 m or more in diameter. That said, the scientific committee responsible for the 30-m telescope has decided that they want to site it in Hawaii because recent legal objections have not been upheld. Nonetheless, astronomers who watch the Sun set over the clouds from the summit of Maunakea, looking across to see Haleakala peeking through on the island of Maui, before embarking on a night's observing are doubly privileged; privileged to be given the chance to work at the best observatory in the world and doubly privileged to have stood on top of this sacred mountain.

While its high altitude sets Maunakea apart as the world's best site for infrared and sub-millimeter astronomy, and hence at the forefront of studying the Chemical Cosmos, it is not perfect. Astronomers are still limited to certain windows, where the absorption of water vapor is less of a problem. In the region of 1 to 2 μm (a micron, or micrometer is 1 millionth of a meter) there are the J and H windows, with the K window starting at 1.9 μm and extending to 2.4 μm. Between 3 and 4 μm are the L windows, and M covers the region in infrared spectrum around 5 μm. These are all windows in what is called the "near" infrared. From about 7–30 μm the mid infrared has a number of windows that can also be used for infrared astronomy depending on the object being studied. But these longer wavelength windows have a nasty habit of closing just when you most need them. At even longer wavelengths around a millimeter or so, the weather and the water vapor in the night sky can be even more frustrating.

How much better, if much more expensive, to rise above it all and get clear of the Earth's troublesome atmosphere altogether!

The commune of Kourou in French Guiana lies on the Atlantic coast of South America, just north of the Equator. Before the arrival of European settlers in the seventeenth century, the commune was home to the Cariban-speaking Galibi people. In the late eighteenth century, thousands of French colonists arrived in the region, enticed across the Atlantic with tales of El Dorado. They did not find it and half died of fever, malaria and other local delicacies; the other half fled back to France. Kourou earned the nickname *l'Enfer Vert*, the Green Hell, on account of its lush but (to Europeans) deadly forests. Off the coast, the Isles de Salut became a prison for the toughest of French criminals, the last of whom left in 1953. Not the most auspicious start for a place that is now synonymous with the most advanced and adventurous scientific expeditions, those setting off for deep space. Yet it was in Kourou that the building of the Centre Spatial Guyanais was started in 1965. Initially a purely French project, the center now launches spacecraft for the European Space Agency, the Russian space agency and NASA.

The Ariane 5 launch vehicle stands nearly 60 m tall and weighs nearly 800 t. As it sets off from the launch pad its rockets burn a mixture of liquid Oxygen and liquid Hydrogen, 155 t in all. These enormous rockets are helped by smaller solid fuel boosters. When the main rockets cut out, just 10 min after ignition, second stage rockets that use Nitrogen Tetroxide to provide the Oxygen needed for burning take over for a similar amount of time. Unlike the first stage, the second stage rockets can be switched off and then restarted should the need arise. From Kourou, the Ariane 5 rocket can lift payloads of between 6 and 10 t, placing them in orbits some 25,000 km into space where they can transfer into an orbit that allows them to hover above the Earth, easily tracked, in a geostationary orbit.

Some time around the end of the decade, if everything goes to plan and enough money can eventually be found on both sides of the Atlantic, the community of scientists who are interested in infrared astronomy may have a dream come true. The designs for the James Webb Space Telescope (JWST), named for a former NASA administrator, give it the largest mirror to collect light from the heavens that has ever flown. It will dwarf the now-aging and much repaired Hubble Space Telescope. At 6½ m in diameter – nearly three times that of Hubble – it will rival the collecting power of all but the largest ground-based telescopes, without having any of the drawbacks of looking through our atmosphere. The main mirror will be kept out of the Sun's rays by a shield the size of a tennis court. Since both of these ambitious structures are far too large to

FIGURE 5.2 An artist's impression of the James Webb Space Telescope in orbit: *credit – NASA*.

fit, as they will operate, into the nose-cone of the Ariane 5 rocket that will lift JWST 1½ million kilometers into space, they will have to unfold like giant umbrellas once the observatory reaches its destination. So, too, will the much smaller, secondary, mirror that has the job of reflecting light captured by the main mirror into the instruments where it can be analyzed. With its impressive toolkit of cameras and spectrometers, JWST will be the latest in the line of infrared astronomical observatories to lift off from Earth (Figure 5.2).

Early attempts to get clear of our atmosphere, in the 1960s and 1970s, made use of sounding rockets and balloons, capable of reaching altitudes of some 80,000 m and opening up much of the infrared and sub-millimeter sky to observation, at least during the relatively short time that they were airborne. In 1974, NASA commissioned a Lockheed C-141 Starlifter airplane as an airborne observatory, which could spend several hours at a time at an altitude of 14,000 m, 10,000 m above Maunakea; they named it for Gerard Kuiper. In over 20 years of service, the Kuiper Airborne Observatory studied the Interstellar Medium, the stars of the Milky Way and galaxies beyond, and the planets of our Solar System. When Supernova 1987A erupted onto the astronomical stage,

FIGURE 5.3 An artist's impression of the Infrared Astronomical Satellite in orbit: *credit – NASA*.

it was there measuring fingerprint spectra from the explosion at wavelengths right across the infrared wavebands (Figure 5.3).

Then in 1983 astronomers got their first really comprehensive picture of the universe at infrared wavelengths when the USA, the UK and the Netherlands launched the Infrared Astronomical Satellite, IRAS. Weighing in at just over a ton, IRAS was launched from the Vandenberg Airforce Base complex in California, on a reliable Delta 3910 rocket, the latest of a family that had been in service in the USA since the 1960s. Compared with JWST, the IRAS mirror was feeble indeed, just 57 cm across. As with all infrared telescopes, the instruments had to be cooled to prevent the heat that they gave off from swamping the signals from space. Limited supplies of liquid Helium coolant meant that the mission could only last 10 months. What IRAS provided, nonetheless, was quite literally a revelation; at infrared wavelengths, invisible to the human eye, the sky was criss-crossed with bright filaments of gas and dust. Familiar constellations, like Orion, lost their hot bright stars like Rigel in a sea of warm glows. Cold cloud upon cold cloud, hot nebula upon hot nebula, showed itself from the former darkness. Stars had vast jets pouring out from them, often in diametrically opposite directions. Bipolar outflows, as they were called, showed how the disks of dust and gas that surround newborn stars and which can give rise to planetary systems, funneled the fast winds that carried the excess gas cloud back out into space.

Among the early achievements of IRAS was to show that while elliptical galaxies were notable by their absence in the infrared, spiral galaxies, like our own Milky Way and the nearby Andromeda Galaxy, and showed up brightly as result of their high gas and dust content. Indeed, warmed by the high rate at which stars were forming, some spiral galaxies were five times as bright in the infrared as they were in the visible part of the spectrum. IRAS found that, like our own Milky Way, the galaxies it found to be the brightest were also full of Carbon Monoxide, CO, with enough star-forming molecular gas to produce billions of stars the size of our Sun. The IRAS spectrometer found traces of Acetylene, Hydrogen Cyanide and Silicon Carbide around known Carbon-rich stars. It provided the first hints that PAHs (Polycyclic Aromatic Hydrocarbons) could potentially exist in such environments.

IRAS mapped the heavens as never before, giving astronomy students subject after subject for their master's dissertations and PhD theses. In total, IRAS detected 350,000 new astronomical objects that had bright infrared emission, increasing astronomers' knowledge of the infrared heavens by some 70%. Literally hundreds of today's infrared astronomers have cut their teeth on IRAS catalogues and data sources. The legacy of IRAS to astronomy is such that more than 11,000 scientific articles, produced at an average rate of over 430 a year since it launched, that owe it at least part of their science, can be found on the Astrophysical Data System hosted by the Harvard Smithsonian Center for Astrophysics. The Chemical Cosmos would never be the same again.

Infrared astronomers had to wait over a decade for the next major space mission, but finally the European Space Agency proved what it could do for astrochemistry with the launch of the Infrared Space Observatory (ISO). The telescope was not much bigger than that of IRAS, 60 cm in diameter, and again the mission time was limited as a result of the amount of Helium coolant that could be launched into space, although ISO lasted 2½ years instead of the expected 18 months. ISO, however, carried spectrometers with good resolutions, ideal for identifying molecule after molecule.

Far above the Earth's wet atmosphere, ISO could map the constellation of Orion in emission from Water molecules, not just in Hydrogen and Carbon Monoxide. It detected unexpected molecules like Hydrogen Fluoride, a corrosive, toxic acid, towards the center of the Milky Way. ISO detected Water on all of the Solar System's giant planets: Jupiter, Saturn, Uranus, and Neptune, as well as Saturn's moon, Titan. Jupiter was also shown to be a guide good to the gas mixture in the early Solar System since its Deuterium to Hydrogen ratio, at 2 parts in 100,000, was exactly as expected from Big Bang cosmology. The Red Rectangle was shown to have grains of the green mineral Olivine in its sooty environment. IRC+10216, the prototypical Carbon star, had a "forest" of spectral lines due to highly excited Hydrogen Cyanide, along with Acetylene. There were ices containing organic molecules such as methanol and methane. Once more, the scientific payback from ISO has been enormous; more than 6,500 scientific papers are listed, at a rate of over 460 each year since launch.

IRAS, ISO and – hopefully – in the not too distant future, JWST are all missions built on the work of ground-based observatories to measure the spectrum of molecules in space across the infrared wavebands. So how, then, do molecules produce the infrared and microwave spectra that the astronomers set out to measure?

Like atoms, molecules produce spectra resulting from changes in the way that the electrons holding them together are arranged. There are *electronic* energy levels and jumps between these levels create distinct fingerprint bands. The most intense of these changes cause lines in the visible or ultraviolet part of the spectrum, although electronic transitions can occur all the way into the microwave. Unlike atoms, however, molecules have spectra that are caused by the motions of the atomic nuclei from which they are made – vibrations and rotations. These motions are, as we might expect, quantized; the vibrational energy and rotational energy of a molecule can only have a certain set of values. Jumps in either vibrational energy or rotational energy, or both at the same time, give rise to individual lines in the spectrum – the spectral fingerprints that allow molecules to be detected and recognized whether they are in a discharge tube in the laboratory, the atmosphere of a distant planet or star, a giant interstellar gas cloud deep in the Milky Way.

Although it is not strictly correct to do so, as a first approximation spectroscopists think of vibrational and rotational motions separately from electronic motions. This is the Born-Oppenheimer approximation, which was first employed in its classical quantum form by Niels Bohr. This gives rise to a natural ordering of energy levels in a molecule; energy jumps in electronic levels are greater than jumps in vibrational levels, which, in turn, are greater than jumps in rotational levels. The simplest molecules to deal with are those that have just two atoms like the diatomic Hydrogen molecule, H_2.

This takes us back to our description of the potential energy surface as a valley between a high cliff and a hill that plateaus out. We will start with the quantized vibrational energy levels. It is as if some ingenious people had set up a system of bridges across the valley at different heights, and changing vibrational level would involve jumping from one bridge to a higher or lower one. The middle of any bridge represents the average distance between the two Hydrogen atoms in the H_2 molecule. They can, however, get closer, represented by the near side of the valley, at the cliff edge of the bridge. Or they can get farther away, over to the far side hill edge of the bridge (Figure 5.4).

The atoms in the H_2 molecule vibrate back and forth along the bridge so that sometimes they are close together, sometimes far apart. This is where the shape of the valley of our potential energy surface comes into play. The depth of the valley tells us about the stability of the molecule. If the valley is very deep, it will take a lot of energy to get up to the plateau where the atoms can drift apart and the molecule will break up; a deep valley therefore corresponds to a stable molecule. On the other hand, molecules that have a potential energy surface with just a shallow valley break up easily.

The width of the valley is also important, however. If the valley is very narrow, with steep sides to both the near side cliff and the far side hill, there is not much room for movement on the bridges. The atoms in the molecule will vibrate back and forth rapidly over a short distance. Conversely, if the valley is wide, there is a lot of room from one side of the bridge to another, and the atoms vibrate back and forth at a more leisurely pace, covering a much greater distance. The shape of the valley then helps to determine the frequency of the vibration, and how that shape changes as you

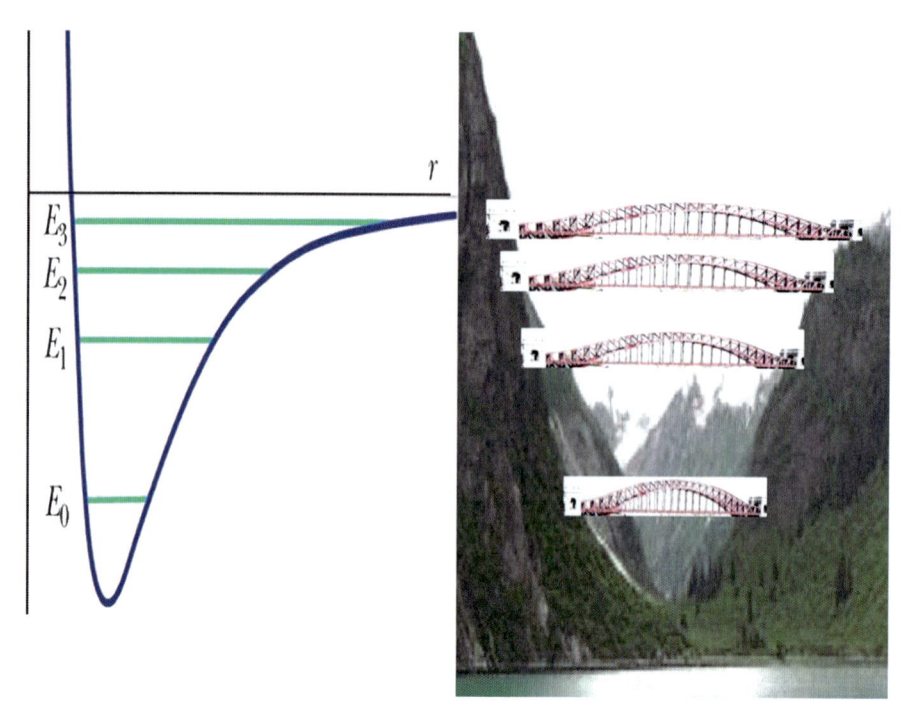

FIGURE 5.4 The vibrational energy levels of a diatomic molecule may be represented by a potential energy "valley" with "bridges" crossing it at various levels: *credit – Steve Miller*.

go up from one bridge to the one above it also determines just how the frequency of the vibration changes.

Vibrations are well known; they did not have to be "invented" just to explain how molecules produce the lines in their spectra. Musicians have long understood the idea of the way a string works on a guitar or violin. Long strings produce low frequency notes, shorter strings high frequency notes. In our valley, similarly, longer bridges correspond to lower frequency vibrations and vice versa.

A stringed musical instrument or the pendulum in a clock behaves as what is called a "simple harmonic oscillator". For a molecule to behave like a simple harmonic oscillator, its potential energy surface valley has to have a very particular shape that is, for obvious reasons, called a harmonic potential. This valley shape results in the vibrational bridges being spaced evenly above one another so that the amount of energy required to jump from bridge to bridge is the same whether the molecule jumps from the first bridge to the second, from the second to the third, or from the third to the fourth.

Molecules, however, do not behave like simple harmonic oscillators, especially as you start to get high up the valley sides where the hill flattens off to a plateau. There, while the individual atoms still cannot get too close together by the nearside cliff, they can get very far apart. As we have also seen, once on the plateau itself, the atoms can get as far away from each other as they like, and the molecule breaks up into separate atoms. As the state of the molecule puts it closer and closer to the plateau, so the jumps between the vibrational bridges get less and less. The molecule vibrates *an*harmonically, and molecules, like our diatomic Hydrogen molecule, vibrate anharmonically – if only slightly so – even when on the lowest of the vibrational energy bridges. Our adventurous guide, H_3^+ is very anharmonic, even at the very lowest vibrational bridges, and that has some unexpected consequences.

To properly understand the spectra of molecules, there is another important concept we must look at briefly – symmetry. The human eye has a real appreciation for symmetry. If we look another person's face, we notice quickly if it is somehow lopsided, if one eye is much bigger than the other, if one ear sticks out while the other is flat against the head. We talk about someone having a "lopsided" grin because when they smile, one side of the mouth goes up much higher than the other. Generally speaking, we have a prejudice in favor of symmetry. People with symmetric features are judged more beautiful or handsome than those whose ears give them the appearance of a taxi with one door open.

Symmetry is a very important idea for Quantum Mechanics, too, including how we think about molecules. The diatomic Hydrogen molecule, H_2, is just two identical Hydrogen atoms joined together by a simple bond. If we swap Hydrogen atom A for Hydrogen atom B, we cannot tell the difference. The diatomic Hydrogen molecule has mirror symmetry; we do not notice the molecule changing when we reflect it in a mirror. We can also rotate it by 180° to swap over the two Hydrogen atoms, and we cannot tell the difference either. If we were, however, to replace one of the Hydrogen atoms by its heavy cousin Deuterium, then all of this symmetry goes away. Reflect the HD molecule in the mirror and Deuterium appears to replace Hydrogen, and vice versa.

Symmetry has very important consequences for the vibrations of a molecule. Joseph Hirschfelder had calculated that our chemical guide, H_3^+ would have its most stable structure as a triangle somewhat between a right-angled triangle and an equilateral triangle. That would mean that it had three different ways, or modes, of vibrating, each with its own particular vibrational frequency, and as Hirschfelder noted, two of those vibrational modes ought to be infrared active and measurable by spectroscopists. When it was found that H_3^+ was a perfectly symmetric, equilateral triangle, that idea changed for a perfect equilateral triangle has only *two* vibrational modes; the two infrared active vibrational modes merged into just one active mode.

Which raises a question: what makes a vibrational mode infrared active or not? Once more the answer is "symmetry", or rather it is a *change* of symmetry. If a molecule has symmetry to start with, then it has to vibrate in such a way that its symmetry changes if it is to produce a strong infrared spectrum. If we take the diatomic Hydrogen molecule, H_2, the two atoms can only vibrate back and forth; they get closer to each other or they get further away, but all of the mirror symmetry and the symmetry of rotating by 180° remains the same. So the symmetry of the molecule does not change, and the vibrations of the diatomic Hydrogen molecule, H_2, have only very weak infrared activity. This is also true of all diatomic molecules in which the two atoms are the same. It is *not* true for diatomics like Hydrogen Chloride, H-Cl, since that molecule never really had any symmetry to start with. When H-Cl jumps from one vibrational bridge to another, it *does* absorb or emit infrared light, producing a strong spectrum.

When our adventurer H_3^+ vibrates, it can do so in two main ways. In the first mode of vibration, each of the three Hydrogen atoms moves in and out in unison as if the molecule is breathing. The equilateral triangle first gets bigger and then smaller, but it remains a perfect equilateral triangle. The symmetry does not change, so this vibrational mode, called v_1 (and pronounced "nu one"), is *not* infrared active. In the second mode of vibration, the motion of the atoms is not perfectly choreographed. One moves in, another moves out, the third stays put, or two atoms move first closer together and then further apart. Whichever way they do it, this vibrational mode causes the perfect equilateral triangle symmetry to be lost and it is strongly infrared active. Jumps in the second, v_2 ("nu two"), mode are accompanied by infrared light being absorbed or emitted, depending whether the molecule is jumping up a bridge, so taking in energy (infrared light), or jumping down from bridge to bridge and giving energy out.

Thus far we have dealt with the motion and energy of the electrons in a molecule, which give rise to our potential energy surface and its cliff, valley, hill, and plateau features. The vibrations we have described form a series of bridges across the valley, pretty evenly spaced to begin with, but then getting closer and closer together as we climb up the valley to the hill top and the plateau that eventually allows the individual atoms to escape from one another and the molecule to break apart. Now we come to rotations, those acrobatic tumblings that molecules have but atoms lack.

It makes sense to leave the rotations until last, since chemists tend to think of a ranking of energy levels between electronic potential energy surfaces, vibrations and rotations, with the associated energy jumps getting small as one goes from electronic jumps to vibrational to rotational. We can also force our analogy with an energy valley just a little bit further, remembering that it is an analogy, not an exact picture of how things are for molecules. If the vibrational energy level states are like bridges between the sides of the potential energy valley, stacked one above the other, then the rotational states associated with any vibrational state are like ladders erected on the bridges (Figure 5.5).

The rotational states are once more quantized, like rungs on the ladder. Higher rungs represent more rotational energy. If a harmonic oscillator is a first approximation to the way in which a molecule vibrates, then the corresponding rotational approximation is that of the rigid rotor. Take the simple diatomic Hydrogen molecule, H_2; in the rigid rotor picture, this looks like two balls on either end of a stick. The individual Hydrogen atoms are the balls, and the bond between them is the stick. The rungs of the ladder are evenly spaced, and on each vibrational bridge, the ladders also have the same distance between the rungs – easy!

Easy? If the diatomic Hydrogen molecule really were this simple, then it would be. But it is not. For starters, the idea that the molecule rotates rigidly becomes less and less tenable as more rotational energy is put into it. Instead, the average distance

FIGURE 5.5 The rotational energy levels of a diatomic molecule may be considered as a system of "ladders" fixed to the vibrational "bridges". Note that the rotational ladder on one bridge can go past the bridge above it, all the way to the top of the potential energy "valley": *credit – Steve Miller*.

between the two atoms increases slightly. This has the effect of making the spacing between the rungs decrease the higher up the ladder you get. Secondly, the average distance between the atoms also increases as you go upwards from vibrational bridge to vibrational bridge, meaning that the distance between the rungs on the ladder on the first bridge is greater than the rungs for the ladder on bridge 2, which – in turn – has a larger rung spacing than for rotational ladder on bridge 3 – and so on up the bridges. Climbing up a rotational ladder does not put you from one bridge to another, however, and it is possible to climb right up from the Ground State to the plateau simply by going up its rotational ladder. (It takes our guide 46 rotational rungs to "get to the top" and break up.)

For our triatomic Hydrogen ion, H_3^+ matters are more complicated even than for its diatomic H_2 cousin. The individual

Hydrogen atoms in our chemical triangle are less tightly held together than for H_2; after all, there are just two electrons to make a bond for three atoms while in H_2 the two electrons only have *two* atoms to hold together. Then there is an additional rotational energy that the molecule picks up when it vibrationally bends out of shape, which can either add to the overall rotational energy or detract from it. It may only have three atoms, but our guide leads a complicated Quantum Mechanical life. That means that you do not have to go far up either the series of vibrational bridges or rotational ladders before trying to predict what energies correspond to what vibrations or rotations, and how the molecule is going to make jumps between them becomes a real handful.

When Joseph Hirschfelder first calculated the vibrations of triangular H_3^+ in 1938, based on *his* potential energy surface, he predicted that the wavelength region in which you would be able to measure the spectrum of the molecule would be around 9 μm. Had Takeshi Oka spent his time fishing in those waters, he would have come back with nothing but tall tales of the one that got away. By the time that Oka was trying to find the H_3^+ infrared spectrum, thankfully, the potential energy surface calculations were much improved; G.D. Carney and R.N. Porter at the New York State University had made calculations that predicted a spectrum at 3.975 μm. Oka and his colleague Jim Watson finally worked out that the correct wavelength was 3.966 μm, just 0.009 μm shorter than the New York calculations. Phew!

The traditional method by which chemists treat a complicated quantum problem is to make a simple calculation – harmonic oscillator for the vibrations, rigid rotor for the rotations. Then allow for a more realistic picture; let the vibrations become a bit anharmonic, the rotations to be a bit less rigid, and allow for a bit of "coupling" between the two. Add a pinch more anharmonicity and a spoonful less rigidity to taste, and pretty soon you have a very tasty chemical calculation. But not for H_3^+. Even with the best ingredients – the best possible potential energy surface to start with – this approach, known as perturbation theory, produces something you can only eat by holding your nose and swallowing very fast. Another approach was needed.

That approach came from Jonathan Tennyson and his colleague at the University of York (in England), Brian Sutcliffe.

They characterized our guide as being a rather floppy, almost spineless, character. It certainly did not have very good posture, for all it was supposed to be a perfect equilateral triangle. A little bit of vibration, a smidgen of rotation, and our guide got very bent out of shape. Therefore they pretty much discarded the idea that H_3^+ had a stable equilibrium shape about which it would vibrate, and they put aside the notion that you should separate vibrations from rotations in the first place. Yes, a molecule had a very definite angular momentum, which Quantum Mechanics recognized as a well-behaved and well-conserved property. But the vibrations could come out of the wash of a calculation that treated each atom as a separate entity whose motion was determined only by the shape of the potential energy surface and where it was in the valley. Luckily for Tennyson's group, by the time his calculations would be needed by astronomers in 1988, Wilfrid Meyer and Peter Botschwina of the University of Bielefeld in Germany had teamed up with Australian Peter Burton to produce a potential energy surface that exactly matched Oka's spectrum.

The approach followed by Jonathan Tennyson and his group allowed them to do two things. Firstly, given Meyer's accurate potential energy surface, they were then able to calculate very accurately the resulting energy bridges and ladders, and the overlaps between them. Secondly, they were to do something the laboratory spectroscopists could not do. They could work out how strong each line, each transition between one ro-vibrational state and another, should be without making any assumptions about what should and should not be allowed. That was important because without knowing how intense a line in the spectrum should be for each molecule that was emitting the line, you could not work out how many molecules were emitting that line. Without knowing the relative strength between lines, you could not work out the temperature either, how hot the molecular hand that left the tell-tale fingerprints was. You would also be left at the mercy of traditional spectroscopy, which predicted that very few types of transitions would be infrared active and thus measurable.

Spectroscopy starting with a harmonic oscillator approach decrees that molecules could only jump from any particular bridge to the bridge above or below it; the molecule could only change its vibrational state by one quantum at a time. Moreover, jumps that involved only a change in the breathing mode of vibration, v_1, which did not change the symmetry of H_3^+, would not be allowed under any circumstances. Tennyson's team showed that the very floppiness of our chemical guide meant that neither of these supposed rules was right for H_3^+. Oka's Ion Factory went on to show that the Tennyson group prediction of strong forbidden v_1 breathing lines was right. Moreover, jumps of two and even three vibrational ladders and other unexpected tricks were all in a day's work for our adventurous little guide.

The unruly behavior of H_3^+ may be on the extreme end of the spectrum. But all molecules have a bit of a wild side, and this poses challenges and opportunities to those wishing to understand the Chemical Cosmos. Infrared spectroscopy sometimes seems to contain just too much information.

6. Branching Out: In the Land of the Giants and Dwarves

When NASA's Galileo spacecraft arrived at Jupiter in December 1995, it was visiting an old friend. The giant of the Solar System (with mass of nearly 2 billion billion billion kilograms it is over 300 times more massive than our home planet Earth) had been visited by four dedicated space missions in the 1970s, and had made a fleeting acquaintance with another spacecraft in 1992. That said, after an absence of several years, you do not expect your visitor to knock on the door, settle down in the living room, and then throw something in your face. But that is precisely what Galileo did to Jupiter; no sooner had it got into position than it fired a probe deep into the planet's atmosphere (Figure 6.1).

One of the key targets of the Galileo mission to Jupiter was to find out just what it is made of and how it is structured. Knowing this is vitally important if we are to understand just what our Solar System was like way back when it formed out of a swirling cloud of gas and dust more than 4½ billion years ago.

We left the Interstellar Medium (ISM) a chemical river of gas and dust, enriched by generations of stars that had lived generating the heavy elements needed for life in the nuclear furnaces at their centers, generously pouring their hearts back out into mix when they finally exploded as supernovae or passed into a more gentle senility as planetary nebulae. For our Solar System to take shape, the river has to divide; part of the ISM has to be shocked, maybe by a nearby supernova explosion, into becoming just that bit denser so that gravity can begin the job of pulling together enough gas and dust to form the Sun, and to have enough left over to make the planets, the surrounding debris of asteroids, comets, and other minor bodies.

As well as contracting under gravity, the gas cloud known as the proto-solar nebula was also spinning. Viewed from "above",

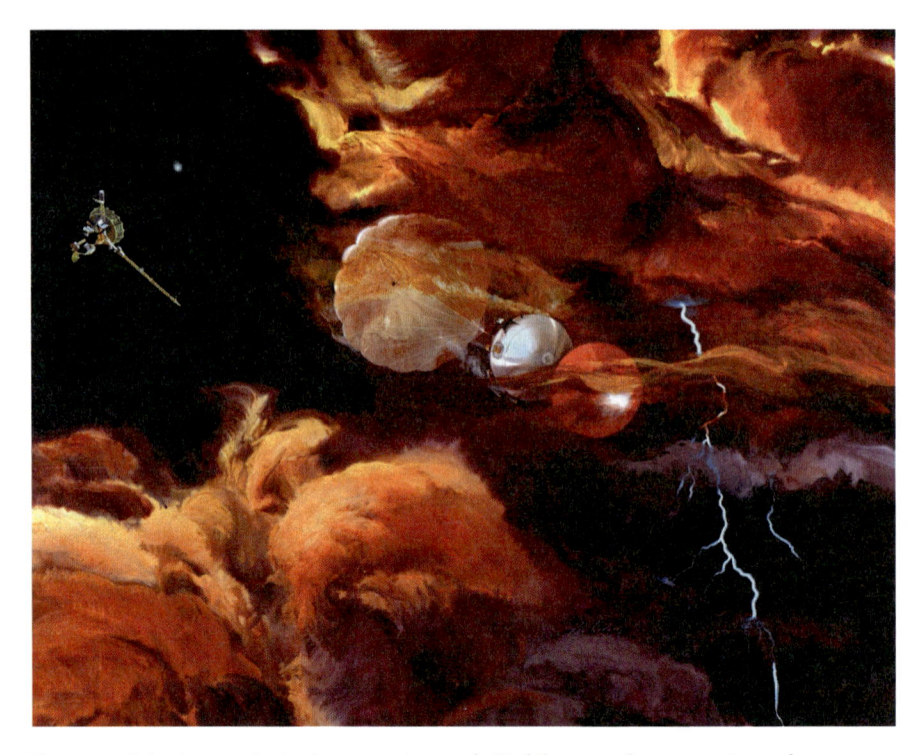

FIGURE 6.1 An artist's impression of Galileo probe entering the atmosphere of Jupiter. In the background the orbiting spacecraft looks on: *credit – NASA*.

from the direction, more or less, that we call north, it span anti-clockwise. The center of the nebula, contracting to form the Sun, would have spun faster and faster, like a ballet dancer drawing in their arms. Today, the Sun spins on its axis about once every twenty five days. It would have ended up spinning much faster, but the angular momentum generated by this spinning was transferred out into the disk of dust and gas surrounding the forming Sun by turbulent waves spreading throughout it and by what is known as "magnetic braking". Here cations like our chemical guide H_3^+ and its deuterated cousin H_2D^+ played a critical role; only ions – positively charged cations or negatively charged electrons – can "feel" a magnetic field and take part in this braking process, helping the Sun lose its angular momentum. While the Sun was forming, its gas/dust disk was itself breaking up to form planets, which could act as repositories for the excess angular momentum. In today's

Solar System, nearly 60% of the total angular momentum is tied up in Jupiter's graceful progress around the Sun. Almost all the rest resides with the other giant planets; Saturn, Uranus and Neptune, and the Sun has less than 4%. This transfer of angular momentum outwards explains why all of the planets and (nearly) all Solar System bodies orbit about the Sun in the same direction.

Our own planet does not give us a very good idea of what the proto-solar nebula, as the gas cloud out of which the Solar System formed is called, was made of. Earth is too small a planet to have held onto an atmosphere that is typical of the proto-solar nebula. The light elements, like Hydrogen and Helium, have already made their escape into interplanetary space. Our planet is also the most geologically active body we know of, weathered by ice and rain, wind and waves, home to vigorous volcanoes, and with enormous restless plates of rock that carry the continents across the surface of the globe. Then there is biology. Much, if not all, of the Earth – in the air, on the land and deep in the sea – is home to living creatures, plants, animals, and more primitive creatures. Throughout billions of years, the activity of all of this life has changed and continues to change the land, the seas and the air.

The Oxygen that we breathe was first introduced into our atmosphere by primitive blue-green algae, called cyanobacteria. These creatures first evolved the capability to carry out photosynthesis, making use of Water, and releasing Oxygen into the atmosphere. Indeed, the fact that we *have* Oxygen in our atmosphere – a highly, corrosive, reactive chemical that ought to be well and truly locked up in the rocks and oceans of Earth – is testimony to the fact that biological activity is still going on; when telescopes, spacecraft and their instruments are sufficiently powerful that they can analyze the chemical composition of Earth-like extrasolar planets, finding Oxygen-rich atmospheres will be key to finding out if there is life outside of the Solar System.

So Earth is a highly processed planet. The Sun tells us much about the elemental abundances found in the proto-solar nebula. But it is hot, with a surface temperature around 6,000° and much hotter than that inside, so that any molecules in the proto-solar nebula have been atomized, with the exception of its cooler regions that show up as sunspots. Here, Water molecules have been found to form and to block out some of the light coming from below.

Jupiter, on the other hand, while not being a pristine record of the proto-solar nebula, is a much better approximation to what the original gas mix must have been like.

Spinning on its axis nearly 2½ as fast as Earth, Jupiter's diameter is about 11 times larger than our planet's. Centrifugal forces mean that this great ball of gas bulges around the middle; from pole-to-pole its diameter is about 6½% less than at the equator. Jupiter has long been known to have a banded appearance (a consequence, in part, of its rapid rotation) with light and dark zones stretching across its disk clearly visible even with quite small telescopes. The Great Red Spot, a vast hurricane so large that it would cover the Earth more than twice over, has been studied for some three hundred years – quite a weather system.

Astronomers only get to see the outer layers of Jupiter. It is hard to get much below the main cloud decks that show up light against a ruddy brown background, and which form at levels where the pressures are about the same as those on the surface of Earth. That means looking inwards through only a few thousand kilometers of jovian gas, not the tens of thousands that would take us all the way to the center of the planet. To understand what lies deep beneath Jupiter's clouds means making sophisticated deductions from spacecraft measurements of the planet's gravity, creating conditions of immensely high pressure in the laboratory, which must exist in Jupiter's interior, and carrying out complicated calculations as what must happen once lab equipment can no longer take the strain.

In terms of chemical elements, Jupiter has much the same general composition as the Sun, mainly Hydrogen, some Helium, and then fractions of a percent of heavier elements. There are differences, which are important, but a Hydrogen/Helium mix is a good starting point. The most likely scenario for the formation of Jupiter (and the other giants) is that large amounts of ice formed up at a distance from the early Sun where the temperature dipped below the freezing point of Water. In a short (by astronomical standards) period of time of 100,000 to a million years, a core of ice and rock built up until it was more than ten times as massive as the Earth. During the next several million years that core had sufficient gravity to attract another 300 Earth-masses-worth of the Hydrogen/Helium rich gas which made up the proto-solar nebula. It was a

Branching Out: In the Land of the Giants and Dwarves   119

race against time, for the young Sun was extremely energetic and was blowing the nebula gas away as fast as the young Jupiter could grab it.

At the center of Jupiter, then, there should be a dense core of ices and rocks. The temperature at the center is over 20,000K. Overlying that, under extreme pressures, many millions of times greater than the pressure we experience on the surface of Earth, is Hydrogen, compressed so much it has become not just a liquid but a liquid metal. As this liquid metal sloshes around the core it generates a magnetic field of enormous proportions, nearly 20,000 times more powerful than Earth's. What is called the atmosphere of Jupiter lies above the liquid Hydrogen layers, with pressures ranging from one million times our surface pressure, well below the clouds, to gas densities corresponding to less than one part in a million millionth at the very top, several thousand kilometers above the cloud decks. Jupiter's gravity is such that the planet is still contracting at a rate of 3 cm every year. Because it radiates out into space nearly 70% more energy than it receives in sunlight, its center is slowly cooling, at a rate of 1K every million years.

There is still some controversy about both the mechanism by which Jupiter formed (some scientists think it formed directly from a denser region of gas within the proto-solar nebula without having to wait for an icy core to come together) and its interior structure. Models of Jupiter's innards require very detailed calculations using Quantum Mechanics, in all its glory, on huge super computers. That said there is much we do understand. According to the most likely models, Helium dissolves in metallic Hydrogen and rains out from high-pressure gas, so there should be less Helium in Jupiter's atmosphere than in the Sun; there is. Helium will also take its inert cousin Neon with it, so Neon should be less abundant than in the Sun; it is. As gas from the proto-solar nebula rained down on the core, it would have heated and evaporated ices containing heavier elements, like Carbon, Nitrogen and Oxygen so there should be more of these elements in the molecules that make up Jupiter's atmosphere. And so there is – for Carbon and Nitrogen.

But not Oxygen. While there is nearly three times as much Carbon-containing material in Jupiter's atmosphere as there is in the Sun, and more than three times as much Nitrogen, only one third of the Oxygen complement of the Sun is found. Oxygen is about ten times less abundant than it should be, compared with Nitrogen and Carbon. So "Go find the Oxygen!" – that was the mission with which Galileo's probe was charged.

It failed.

This is not to criticize the poor old probe. It was set free from the main Galileo spacecraft some 5 months before final arrival at the giant planet and left to its own devices. By the time the probe made its suicidal descent into Jupiter's atmosphere on December 7, 1995, it had braved the planet's vast radiation belts, suffering doses that would have fried all but the most hardened electronics. For 58 courageous minutes it told astronomers back at NASA's Jet Propulsion Laboratory in Pasadena what it found. Eventually, temperatures greater than boiling water and pressures equivalent to diving to more than 200 m below the ocean surface burned and crushed the life out of it. The problem for the scientists hanging with bated breath on the Galileo probe's every word was that it did not say what they wanted to hear.

In a Hydrogen-rich atmosphere like Jupiter's most of the heavier elements are bound up as Hydrides. Nitrogen forms Ammonia, NH_3; Carbon forms HydroCarbons like Methane; CH_4. Sulfur produces the evil-smelling Hydrogen Sulfide; H_2S and Phosphorus forms Phosphine; PH_3, in part responsible for the colors of Jupiter's bands. And Oxygen ought to form Water, H_2O. The probe passed through high altitude Hydrocarbons and found a cloud layer of Ammonia ice, and it probably found traces of complex Ammonium Hydrogen Sulfide clouds as well. There was not a sign of the Water clouds, however, although the probe had penetrated much deeper than should have been needed to find it. Fortunate then that JPL's ground-based astronomers were also monitoring Jupiter at the time the probe entered the atmosphere. It was they who provided the answer to the mystery; the probe, by an unlucky accident, had chosen a particularly dry part of Jupiter's atmosphere to enter, where cooler air from high up above the clouds was pouring back down creating a Water-free zone. The probe's reputation was saved, but the missing Water remains missing, an embarrassment for planetary scientists.

While Jupiter's atmosphere confirms what the *elemental* make up of the proto-solar nebula was like, with the exception

Branching Out: In the Land of the Giants and Dwarves   121

of the missing Oxygen and relatively low abundances of Helium (and Neon), the Chemical Cosmos is characterized by continual change. High above the planet's cloud decks, in the stratosphere, sunlight is doing what starlight always does, which is to change the chemistry on a daily basis. In some ways, Jupiter's atmosphere gives us a speeded-up version of what is happening in the clouds of the Interstellar Medium. It is closer to its source of radiation – our Sun – than most interstellar clouds and that drives chemical reactions faster. The stratosphere is also much denser than the ISM; pressures there vary from about one third of sea level pressure on Earth, at the bottom of the stratosphere, to one millionth of the pressure we experience, at the top. Even at that low pressure, a microbar or so, there are about 100 billion billion molecules in every cubic meter of air, billions of times as many as in the dense interstellar clouds. Therefore there is a much greater chance of an atom or molecule bumping into another one to create a new chemical compound; the time for a reaction to take place is consequently much shorter making Jupiter's atmosphere look quite like the ISM on overdrive, a nice test-bed for interstellar chemistry that would otherwise take too long for us to study.

Apart from Hydrogen and Helium, Methane is the most abundant gas in the stratosphere. Sunlight will just not leave Methane alone, however; high-energy photons have the power to kick a Hydrogen atom out of its cozy home, creating CH_3 from CH_4. This *Methyl radical* is – as the name suggests – a bit of a firebrand. Pretty soon, and as in the ISM, two of these radicals team up to form Ethane, the stuff of lighter fuel, which has the chemical formula C_2H_6 and can be thought of as two Methyl radicals holding hands through their Carbon atoms, H_3C-CH_3. Yet even the radicals can be affected by sunlight, losing more and more Hydrogen atoms until just one is left, CH. These Hydrogen-hungry creatures are let loose on the peaceful Methane molecules, forming compounds like Ethylene and, with the help of further sunlight, Acetylene the blow-torch gas. All in all the effect of sunlight on Methane gives rise to a spider's web of Carbon chemistry involving more than 40 compounds, including the formation of the Carbon-ring compound Benzene and its cousins.

Huge molecules, like the ones that eventually give rise to large grains of Carbon in the darkest reaches of the ISM, do not form easily in Jupiter's atmosphere. There is just too much going on, too much sunlight to break apart the chemical bonds that have laboriously been put together. At the poles, however, where there is much less sunlight, tarry hazes form up, the kinds of molecules that smokers seem to enjoy. No one quite knows what they are made of, but there may be some molecules not that dissimilar from those that eventually formed the prebiotic soup from which life evolved on Earth.

Higher still in Jupiter's atmosphere, the air is no longer nicely mixed up by upward and downward currents. Instead, as the temperature rises from just 100 or so degrees above absolute zero to well over 1,000K, chemicals settle out according to their own weight like colored sands falling to the bottom of a jar of sea water. As we reach from a few hundred to a few thousand kilometers above the cloud tops, air pressure drops from one millionth of our sea-level pressure to fractions of a billionth of a bar. Pretty soon, only Hydrogen and Helium are left. This thin gruel is bathed in ultraviolet sunlight, and – around the northern and southern poles – bombarded by energetic particles. It is the ideal place for ionization to take place; it is the ideal place for our chemical guide to form. From its vantage point high above the atmosphere of Jupiter, H_3^+ oversees this microcosm of our Chemical Cosmos. Bright H_3^+ aurorae shine out a thousand times brighter than Earth's Northern and Southern Lights. Across the whole planet, there is a warm glow of H_3^+ emission (Figure 6.2).

So Jupiter's turbulent atmosphere tells us much about the gas in the proto-solar nebula, but what about the ices and grains that must also have been present?

The emerging Solar System was a very violent place. Seared by a Sun that was many times more energetic, the bits and pieces that were to form the planets – all the way from small rocks to proto-planets bigger than the Moon – crashed back and forth, hitting one another, sticking together or breaking each other apart for some 700 million years. A little less than 4 billion years ago, the Solar System eventually "settled down" to more or less what it is today. So the Solar System we enjoy is most definitely a neighborhood that has cleaned up its act and become respectable. Nonetheless there are still some of the rougher, old inhabitants around, just to keep us from becoming too self-satisfied and complacent, still

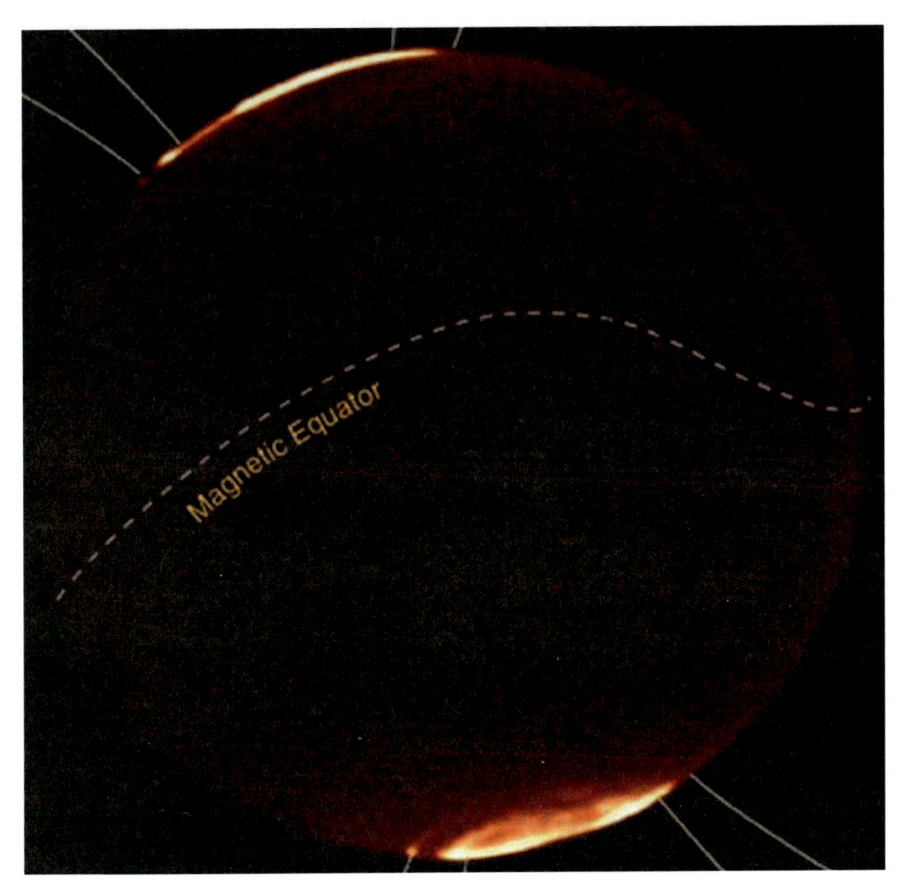

Branching Out: In the Land of the Giants and Dwarves 123

FIGURE 6.2 An image of Jupiter taken at infrared wavelengths sensitive to H_3^+. The image shows the southern aurora clearly, as the planet is tilted in our line of sight, and the northern aurora peaking just over the top of the planet. As well as a warm H_3^+ glow across the planet, a magnetic footprint of Io and its orbit can also be seen. The magnetic equator (dashed line) and field lines (solid) connecting to the inner aurora (nearest the poles) and Io are indicated: *credit – Jack Connerney, Takehiko Satoh, NASA Infrared Telescope Facility.*

violent episodes; a comet or asteroid will crash into a planet with greater regularity than we might imagine.

Just after 1 a.m. on February 8, 1969, the night sky over Allende, in the Mexican state of Chihuahua, was lit up by a giant fireball. Approaching from the southwest, a huge rock, the size of an automobile was ploughing into the atmosphere at more than 15 km/s. Unbearable frictional forces were heating it as it made its last journey from interplanetary space down to our planet; so great

124   The Chemical Cosmos

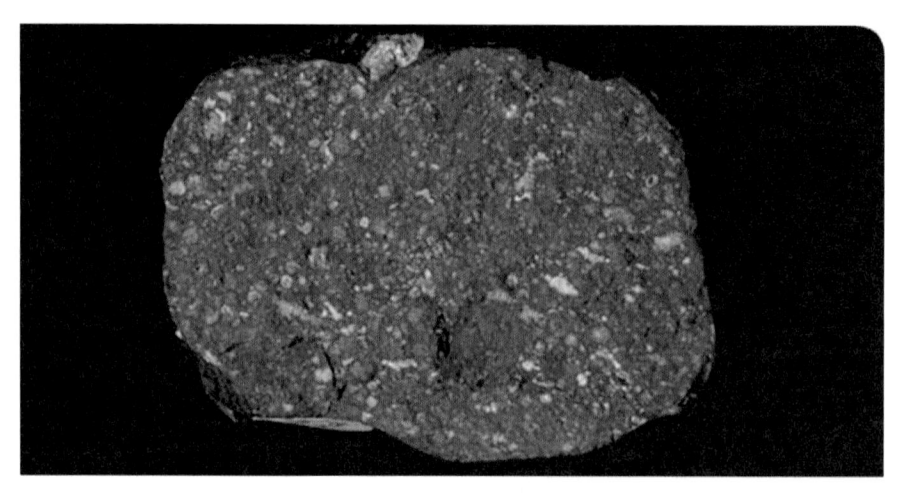

FIGURE 6.3 A fragment of the Allende meteorite held at the Natural History Museum in London: *credit – Natural History Museum, London.*

were the forces that the rock shattered. Some of the pieces that reached Earth were just small pebbles, a gram or so in weight and others were over 100 kg. Being hit by either would have proved painful, given how fast they were traveling. Over the next quarter of a century, either by accident or design, several tons of what is collectively known as the Allende Meteorite have been found and handed over to museums, universities and other places of learning. Allende is probably the best studied meteorite in the world.

Meteorites, and the comets and asteroids that go with them, are the leftovers from the violent formation of the Solar System. Too small to form into planets, often banished to the outer edges of interplanetary space, comets preserve a lot of the original ices in the proto-solar nebula. Asteroids, those in the Main Belt between Mars and Jupiter especially, were never given the chance to get together into planet-sized bodies, although one asteroid called Ceres got close and is now classed – controversially, along with Pluto – as a Dwarf Planet. Meteorites are chips of the old block, off colliding asteroids, off the Moon, off Mars, off anything.

Allende really is a relic (Figure 6.3). It is one of the oldest "bits" of the Solar System that we can actually get our hands on. Indeed, it has been dated to within a couple of million years *before* the official formation of the Solar System, and along with similar meteorites enables us to date that formation with great precision to just over 4,567 million years before present. Within Allende, rare forms of Calcium and Barium, and the even rarer element Neodymium, support the idea that the Solar System formed as a shock wave from an exploding supernova passed through the nearby interstellar gas cloud. Rare forms of Beryllium and Sulfur, itself a leftover from radioactive Chlorine, show that the early Sun, in its turn, sent out its own shock wave of radioactive particles.

Much of Allende is formed from the rocky minerals Olivine and Pyroxene, rich in the metals Magnesium, Iron, Aluminum, and Titanium, and in glassy Silicates. The grains of the proto-solar nebula must have been composed of such materials, their surfaces forming workbenches for chemical reactions to take place in the cold depths of the ISM, inaccessible to all but the most penetrating radiation. Almost certainly, these grains were processed – melted and reformed – as the supernova shockwave passed through them; only minute diamonds may have survived the supernova explosion. But the grains formed up again before the actual birth of the Solar System itself. They had to; grains are vital, along with molecules, to cool down the proto-solar nebula sufficiently for the young Sun to condense and become a Hydrogen-burning star.

Allende is also a storehouse of Carbon and is known as a carbonaceous chondrite. The Carbon in Allende is found as the element, as crystals of diamond and as sheets of graphite. Allende houses the third form of Carbon, football-shaped Fullerene molecules, some combined with small amounts of Hydrogen. The simplest Fullerene molecules have cage-like structures containing 60 or 70 Carbon atoms. In Allende, however, as well as C_{60} and C_{70}, there are cages that have between 100 and 400 Carbon atoms, perhaps an entirely new range of aromatic molecules. Allende certainly contains a rich soup of organic compounds, including Amino Acids and Polycyclic Aromatic Hydrocarbons (PAHs), testimony to the complex chemistry that evolved on the dust grains of the proto-solar nebula.

1969 was a good year for meteorites. As well as the fall of Allende in the February of that year, the month of September witnessed a meteorite of perhaps even greater significance. The Murchison meteorite, which landed near the town of the same name in the Australian state of Victoria on the morning of September 28th, is another carbonaceous chondrite like Allende, rich in organic molecules. Like Allende, Murchison has been extensively studied, its composition probed to see just how far the chemistry of the Solar System progressed in those early days of the proto-solar nebula. A long way is the answer; as well as a healthy smattering of Amino Acids, including terrestrial and non-terrestrial types, Murchison has recently been found to contain "nucleobases".

In 2008, Zita Martins from the Leiden Institute of Chemistry in the Netherlands and an international team carefully analyzed material from deep inside the Murchison meteorite to ensure that any contamination from Earth could be accounted for and eliminated from their results. They were able to demonstrate that the compounds Uracil and Xanthine, which showed up with high levels of the heavy Carbon isotope, $_6^{13}C$, compared with normal Carbon-12, could not have come from contamination on Earth. Uracil is a modern nucleobase known as a Pyrimidine, a component of the genetic material of viruses, RiboNucleic Acid or RNA. Pyrimidines have an aromatic ring of six atoms, four of which are Carbon and two Nitrogen. Xanthine is a Purine, which has a Nitrogen-containing five-atom ring fused into the Pyrimidine ring, and may have been a nucleobase in the past, although it is nowadays less biologically important.

For many years, planetary scientists had been pointing to the possibility that enormous amounts of organic material had rained down on Earth as part of the deluge of small chunks of rock and ice that occurred during the first half-a-billion or so years of the Solar System, a period of heavy bombardments as our planetary system finally took shape. The famous American astronomer Carl Sagan calculated that this Carbon-rich "rain" could have amounted to a million tons a year, possibly seeding the early Earth with life-giving organic materials. What Martins' team had shown is that this rain could have included chemicals, such as Uracil and Xanthine, capable of what they called "self-association", of taking the first steps towards forming self-replicating molecules.

Most meteorites are chips off asteroids, although there are some that come from the Moon and from the planet Mars. They tell us much about the rocky part of the grains that were present in the proto-solar nebula, and what chemicals could be included in those grains. In our pinball machine of a Solar System, Earth is also visited by other bodies, bodies that tell us much about the icy part of the nebula – comets.

Comets have been objects of dread since humans began to record their thoughts and emotions. Comets (their name means "hairy") with their long bright tails streaking across the night sky seemed to have come out of nowhere to disrupt the tranquil certainty of the heavens. They could only be up to no good; something momentous was about to happen. Famously, the March 1066 appearance of Halley's Comet is immortalized on the Bayeux Tapestry, which celebrates the defeat of the Saxon king Alfred by the Franco-Viking William Duke of Normandy, at the Battle of Senlac Hill near Hastings in southern England, on October 14th of that fateful year. Although comets have been observed and studied for as long as human records exist, and almost certainly before that too, understanding just what they are is a relatively recent achievement. For a start, where do they come from? Do comets originate in our Solar System, or are they somehow captured from interstellar space? And what are they made of; are comets balls of gas, loosely bound grains, or do they have a more solid construction?

Fred Whipple of the Harvard College Observatory (he later became director of the Smithsonian Astronomical Observatory) had been studying comets since the 1930s. From 1950 onwards, he published a series of papers that pretty much laid the basis for our current understanding of the composition of comets. Whipple's suggestion was that comets had a fairly robust nucleus, a few to a few tens of kilometers in diameter, composed of ices such as: Water ice, solid H_2O, Ammonia ice, solid NH_3, and icy Methane, solid CH_4, as well as solids of either Carbon Monoxide, CO, or Carbon Dioxide, CO_2, or both. At the time Whipple put forward these ideas, none of these compounds had been detected in comets, but his model was able to account for the gases that *had* been seen in cometary spectra. Whipple's analysis of the way comets traveled past the Sun and planets favored them having formed in the Solar System, rather than in interstellar space. For Methane to have solidified required the ices to have formed at very low temperatures, a few tens of degrees above Absolute Zero, and that also meant that comets had to have formed from the icy grains of the proto-solar nebula far away from the boiling effects of the early, vigorous Sun, tens of thousands of times as far away as the Earth, out in what came to be called the Opik-Oort Cloud named for the two astronomers who independently proposed this structure.

Whipple called his structure the "icy conglomerate" model, but it soon became known, less respectfully, as the "dirty snowball". While measurements from the ground increasingly supported the dirty snowball structure, it was not until the mid-1980s, some 30 years after the initial proposal, that astronomers obtained pictures of the cometary nucleus that clearly illustrated that Whipple was right. In 1986, that most famous of comets, Halley, flew through the inner Solar System close to the Earth. That March, the twin Soviet Vega spacecrafts, heading off from Venus into interplanetary space, swung bravely by to take a look; Vega 1 approached to within 9,000 km of the comet and Vega 2 got even closer, a mere 8,000 km away. The hundreds of images that the Vegas sent back to Earth showed a dark, but remarkably warm, nucleus, with two bright gas jets erupting from it. Some of the material boiling off Halley was surprisingly like the carbonaceous chondrites found in the Allende and Murchison meteorites, and they were simple molecules trapped in a matrix of Water ice. Damage to the craft caused by the bombardment of particles from Halley's tail caused Vega 1 to lose 40% of its power; Vega 2 lost 80%. Clearly getting to close to a comet was a dangerous business.

The Vega spacecrafts helped to pioneer the way for an even more adventurous project, the European Space Agency's Giotto space probe, named for the medieval Italian painter Giotto di Bondone, who had included his own observations of Halley's 1301 flyby of Earth in his version of the Adoration of the Magi. Protected by a dust shield designed by Whipple himself, Giotto was to get within 600 km of Halley's Comet on March 14, 1986, just a few days after the Vega flybys. It was a rocky ride. Hammered by debris flying off the nucleus, Giotto was sent spinning such that its instruments were no longer protected by the dust shield. The multi-color camera was smashed just as it got closest. British Prime Minister Margaret Thatcher watched the encounter live on television, along with millions of her fellow citizens. The loss of picture when the camera went down is said to have left her "unimpressed" with the whole affair. Fortunately, there was much for the Giotto scientists to get their teeth into, whatever the Iron Lady's disappointment.

Giotto showed the nucleus of Halley to be a "peanut-shaped" dust covered body, "blacker than coal", some 15 km long and between 7 and 10 km wide. Eighty percent of the gas ejected into the tail was Water, 10% was Carbon Monoxide, CO, and about 2½% was a mixture of Methane and Ammonia. Whipple's ices were all there, or at least the gases derived from them were, but the snowball – or at least its surface – was "dirtier" even than he had expected. So was it possible to get below the surface? Two missions early in the new century embarked to find out.

The chorus in Joni Mitchell's tribute to the 1969 Woodstock Rock Festival has the line "we are stardust, million year old carbon". In February 1999, NASA's Stardust Mission to Comet Wild-2 took up the theme, if not the chronology, of Mitchell's haunting refrain. Following Giotto's 1986 trailblazing journey to Comet Halley, Stardust set off for a close encounter with Comet Wild-2. It arrived at its destination on January 2, 2004, having looped twice round the Sun to pick up enough energy to get out to the comet. The spacecraft got within 240 km of the comet. Its cameras imaged over 20 jets of gas and dust erupting from the nucleus, icy volcanism on a grand scale. But Stardust's mission, unlike that of Giotto, was not simply to take pictures and make measurements of the space environment of the comet; Stardust was to bring a little bit of Wild-2 back to Earth, a little bit of stardust.

Just over 2 years later, after another trip around the Sun, Stardust came back to Earth with a bump. As it crashed unceremoniously into the Great Salt Lake desert in Utah, there were fears that its precious cometary samples would be contaminated, mixed with earthly sand. Fortunately, those fears proved unfounded. Over a thousand particles from Wild-2's dusty jets, with sizes between a few millionths of a meter to one third of a millimeter had been scooped up into the spacecraft's aerogel collector, a porous sheet made up of nanometer-sized glass hairs. Over 200 scientists from around the world, led by Don Brownlee of the University of Washington in Seattle, set to work analyzing the priceless samples. *Science* devoted its December 15, 2006, edition to their findings.

Grains of Olivine and Pyroxine were, as expected, common. The grains were full of metals and metal compounds such as Magnesium, Aluminum and Calcium, as well as Iron, Chromium and Nickel, produced by a nearby star going supernova. Nonetheless, there were some major surprises. Although some of the grains found in Stardust's collection were pre-solar, direct from the Interstellar Medium out of which the nebula condensed, most were not.

They testified to being melted and reformed close to the early Sun and then sent out into the outer reaches of the Solar System where ices could form up and build the nuclei of comets. The organic, Carbon-bearing, molecules returned by Stardust included many of the usual suspects including PAHs with anything between two and six rings of Carbon fused together. These molecules told a more ambiguous story than the grains, with a mixed parentage from both the dense ISM and proto-solar nebula. If Wild-2 were typical, the early Solar System was clearly a thoroughly mixed up place.

Nineteen months after Stardust's encounter with Wild-2, NASA's Deep Impact mission reached Comet Tempel-1, on Independence Day, 2005. Rather than wait for the spacecraft to be bombarded by the dust and debris thrown off by the comet, NASA decided to get their retaliation in first. In a "life-imitating-art" gesture, the real mission fired a 370-kilogram block of Copper metal at the nucleus of Tempel-1. Unlike the Bruce Willis version, however, Tempel-1 was to carry on pretty much unscathed except for a new crater a couple of hundred meters across. As well as the cameras onboard the Deep Impact spacecraft itself, the Hubble Space Telescope watched to see what would happen, along with another NASA mission called Spitzer (of which, more later). On the ground, every observatory that could focused its telescopes on the comet, linked by a video conferencing system so that observations could be reported across the globe as they were being made; Impact Time was perfectly set for the telescopes on Maunakea. The Copper probe was due to hit on July 4th at 05:45 Universal Time, and by the time the light from impact reached Earth, it would be just past 10 min to 8 p.m. on the evening of July 3rd in Hawaii.

Deep Impact was quite a gamble, and, as such, it engendered quite a bit of nervousness. Project leader Mike A'Hearn had already run the gauntlet of NASA review committees that had threatened to cancel the mission as budgets tightened, costs overran and computers refused to perform up to par. As the spacecraft approached Tempel-1 it had to release the impactor just right; for its final day, the projectile would have to travel the last 800,000 km to the comet under its own steam, making small corrections to its trajectory using its own onboard rockets. There was little room for error. As impact time approached the atmosphere at NASA's

Jet Propulsion Lab in Pasadena grew quieter and tenser: would it be a hit or a miss? While JPL waited for images to be relayed down from the Deep Impact spacecraft, it fell to the United Kingdom InfraRed Telescope (UKIRT) on Maunakea to report to the world that it was a hit.

At 05:50:52 Universal Time on July 4th, UKIRT's guide camera saw the comet start to brighten. The observation went global via the video network. During the next hour, the brightness of Tempel-1 increased tenfold as gas, dust and ice crystals poured off the impact site, reflecting sunlight back to Earth. The spectrum of Water, shocked to high temperatures by the impact and sudden exposure to the Sun's ultraviolet rays, shone brightly out from the newly excavated materials. The Deep Impact spacecraft's own cameras showed a flash as the Copper probe hit, followed by a plume of vaporized ices and a cloud of material ejected from the comet's surface. Across the summit of Maunakea from UKIRT, the giant Keck Telescope was measuring the spectrum of all the molecules that were coming from the impact.

As well as Water, Carbon Monoxide and Methane, Keck identified Ethane, Acetylene, Formaldehyde and Methanol in the impact ejecta; Ethane buried below the comet's surface was much more abundant in the post-impact spectrum. Hydrogen Cyanide was also there, lethal to humans but the feedstock of Amino Acids and proteins. Other telescopes picked up the tell-tale fingerprints of Carbon Dioxide and PAHs and – surprisingly – Carbonates, the chemicals of bones and shells: surprisingly because Carbonates need *liquid* water to form, but all the Water in comets should be solid ice or vapor, not liquid. Answer one puzzle, and another rears its head!

This series of "hit-and-run" flybys initiated in the 1980s by the Vega and Giotto spacecraft have whetted an insatiable appetite for information from comets and their immediate environs. On March 2, 2004, the European Space Agency's Rosetta mission set off on a 10-year journey out beyond the orbit of Jupiter, nearly 800 million kilometers from the Sun, to rendezvous with Comet Churyumov-Gerasimenko (Figure 6.4). Rosetta is not going all that way just to say "hi and goodbye"; once at the comet in 2014 it will orbit it for 2 years as Churyumov-Gerasimenko plunges in towards the Sun, measuring physical and chemical properties

FIGURE 6.4 The launch of ESA's Rosetta mission on March 2, 2004: *credit – the European Space Agency.*

as they change with ever-increasing doses of sunlight and high-energy particles.

Rosetta will also drop a small lander onto the surface of Churyumov-Gerasimenko, sounding the nucleus to find out how dense or fragile it is, analyzing the gases given off before the Sun's ultraviolet radiation has a chance to work any changes. With its suite of imaging and probing instruments, the Rosetta *Mission* hopes to be the Rosetta *Stone* of cometary science, translating the languages of the ISM, the proto-solar nebula and the early Solar System. Maybe Rosetta will even find traces of the liquid water that the Carbonates detected by Deep Impact require.

In all, some 25 gas molecules have now been detected coming from comets, including common ones such as Water and Carbon Dioxide, and rarer gases such as Hydrogen Sulfide and Carbon

Disulfide, Sulfur analogues of the previous two. This is less than in the ISM as a whole, although, as we have seen, that term covers a multitude of environments, but still shows an impressive range of chemistry. The Stardust grains – like the Carbonaceous Chondrite meteorites - are a rich source of organic materials, and many others.

The chemistry of the entire collection of dwarf bodies (comets, asteroids and meteorites) and that of the planets were both separate and yet intertwined in the early Solar System. *Separate*, because the meteorites, the asteroids from which most of them are derived, and the comets we can study today have lived independent lives for some 4½ billion years; these dwarves are the lucky ones, the ones that did not end up plummeting downwards in the gravitational field of one of the major planets. *Intertwined*, because the chemical make-up of the atmospheres and surfaces of planets, if not their deep interiors, has clearly been affected in many ways by the impacts of comets and asteroids. Indeed, much of the Water we have on Earth was delivered by comets hurled in our direction by the newly formed Jupiter and its giant siblings, according to the most probable theories of how the Solar System formed.

Theories are all very well, but seeing is believing. There is nothing like an eye-witness account to carry conviction. If the dinosaurs could just have left a record – a video recording, say – of what they saw and felt when the 10-km-wide comet or asteroid that wiped them off the face of the Earth hit 65 million years ago, how priceless that would be for today's scientists who want to really understand just how mixed up bits and pieces of the Solar System can get. It would take great presence of mind, however, to think of the curiosity of future species when your own was on the point of extinction. Would we do the same for the descendants of the cockroaches who might inherit our planet should we be heading the same way as the dinosaurs? Probably not. But what if you could watch a collision of cosmic proportions happening from a safe, but ringside, seat?

In the early Solar System, Jupiter was a bit of a bully, hurling comets and other bits of rock at the little guys, the inner planets Mars, Earth, Venus, and Mercury. It was such a bully that it kept the rocks of the Main Asteroid Belt from forming their own planet, a cosmic version of "divide and rule" practiced so effectively by empire-builders throughout history. Nowadays Jupiter is a law-abiding citizen, and in the present Solar System it plays the role of border patrol. Comets trying to get in towards the Sun have to get past the gravitational field of the giant planet. Many do not make it, and are put into orbits that keep them relatively far from the Sun. Those even less fortunate may be captured into orbit around Jupiter itself and that can be fatal; David and Goliath notwithstanding, when the dwarf takes on the giant there is only one winner.

Sometime between the two World Wars, a pretty ordinary comet, with an icy nucleus like Halley's or a bit smaller, tried its luck sneaking in towards the Sun. But Jupiter was on the lookout and the unfortunate comet found itself trapped into orbiting the giant planet instead. This was an orbit that had it sometimes a comfortable 50 million kilometers out into space, but at other times it was hazardously close. In 1992, it went skimming just 40,000 km above the cloud decks. Its luck ran out. At this perilous proximity, Jupiter's gravity was enough to rip the nucleus into several fragments.

On March 24, 1993, comet-watcher Carolyn Shoemaker of the US Geological Survey was examining some photographic plates that she and her husband Eugene had taken, along with their colleague David Levy, at the Palomar Observatory in California the previous night. The weather had not been kind and Shoemaker had been quite reluctant to look through what she expected to be an unpromising collection of photos. To her enormous surprise she discovered what she thought looked like a squashed comet: "It was as if someone had stepped on it," she remarked. The news of this rare and very odd object was flashed to astronomers worldwide through the International Astronomical Union's circular email alert. Pretty soon several observatories had confirmed Shoemaker's discovery, and it became clear that this was a comet in orbit around Jupiter, not the Sun.

The Hubble Space Telescope, its original faulty vision recently corrected thanks to a dedicated Shuttle repair mission, turned Jupiter-wards to get the clearest view possible. Hubble showed Comet Shoemaker-Levy-9 (the ninth to be discovered by this prolific team of comet hunters) to be made up of some 20 individual nuclei following each other around the planet like a train and carriages. The break-up of the original nucleus had created enough dust and ice crystals for the comet to be discovered from Earth; Hubble's instruments had sufficient penetration to get beyond the dusty glare to see what lay beneath. Moreover, SL9 – as the comet-lets were collectively named – was on collision course with Jupiter.

A chance, for the first time, to watch a collision between two solar-system bodies of the kind that was common during the formation of the planets was a temptation no astronomer could resist. Observatories around the world went onto alert; impacts would occur at more or less regular intervals between July 16 and 22, 1994. Preparations were made, observing teams were set up, and anticipation rippled far out beyond the scientists immediately involved. The media were highly excited, and journalists and broadcasters welcomed the fact that they could prepare well in advance for events that would occur during the political doldrums of the summer months. Some UK newspapers and broadcasts even carried prophecies from a "plain-clothes nun" by the name of Sister Sofia, who predicted that Jupiter would explode when hit by SL9 – a warning to world leaders to mend their wicked ways. In reply, veteran British astro-broadcaster Patrick Moore pointed out that SL9 would have the same impact on Jupiter, as a whole, as would a pea-shooter fired at a charging rhinoceros.

In the event, no one apart from the doom-sayers was disappointed. The media got a great show, just when they needed it. Public hits on the special SL9 website hit an all-time record high, popular science authors got good sales for books on comets, and the hundreds of astronomers, including many amateurs, watching world-wide had more data than they could handle. Ten years after the event over 300 papers about SL9 had been published in the scientific literature, and the current score on the NASA/Smithsonian database is some 1,200.

Smashing into Jupiter at over 60 km/s, the fragments hit with a flash, a bang and a wallop. There was a flash as the nuclear fragments first hit the top of the atmosphere, heating like meteorites streaking across the night sky. There was bang as, close to Jupiter's cloud decks, the friction got just too much and they exploded, firing a jet of hot gas and dust – heated to tens of thousands of degrees – thousands of kilometers back through the tunnel they had forged on the way down. And there was a wallop as the dust and gases fired into space smashed back down onto the planet, for the main part unable to escape its enormous gravitational pull. Each impact left its scar (Figure 6.5).

136 The Chemical Cosmos

FIGURE 6.5 An infrared image of Jupiter after the impacts of Comet Shoemaker-Levy 9. A fragment is seen hitting on the dawn (left) limb of the planet. Elsewhere, impact scars from previous impacts can be seen: *credit – the Max Planck Institute, Germany.*

The biggest fragments created explosions of several million megatons, dwarfing the world's nuclear arsenal by a factor of 100 or more. They left ring-shaped impact scars that were thousands of kilometers across; Earth would have fitted into the largest. At the end of Impact Week, the belt around Latitude South 45° on Jupiter was a mass of dark rings and spots. Some of the impactors had a nice area all of their own, while others had fallen almost right on top of one another, leaving scar upon scar.

The total destruction of SL9 gave astronomers some key insights into the very depths of comets. For a start, the ices of the nucleus could not have been too tightly held together. Whipple's dirty snowball was more like a rubble pile of smaller snowballs that had teamed up under their own gravitational attraction. No wonder the shape of some cometary nuclei was so odd. Made of Water and other ices, SL9 added large amounts of Oxygen, Carbon and Nitrogen to Jupiter's atmospheric mix. It was also clear that comets were delivering significant amounts of metals like Sodium, Magnesium and Calcium, and the supernova products Iron, Chromium and Manganese.

The impacts altered Jupiter's atmosphere in other ways, too. Plumes of hot Methane and Ammonia were fired up to the highest levels where normally they cannot reach. Methane was even seen traveling so rapidly that it would escape even Jupiter's gravitational clutches; small amounts of the atmosphere were being hammered off as a result of the collisions. In the early Solar System, impacts must have had this effect many times over, often boiling off more atmosphere than they delivered. Chemicals new to Jupiter were formed in the extreme heat of the final explosions and Hydrogen Cyanide was one of these. Previously undetected anywhere on the planet, over the years following SL9 it spread right across the sphere. Then there were the scar sites. Their dark hues had to be the result of the formation of organic, tar-like materials. Like Hydrogen Cyanide, these too spread across the planet, eventually settling downward below the cloud decks.

No part of Jupiter's atmosphere, from the cloud decks upwards, was unaffected. From its lighthouse vantage point high in Jupiter's polar atmosphere, our chemical guide H_3^+ initially oversaw proceedings relatively untroubled by the mayhem going on below. Its bright infrared auroral light shone out, a constant reminder that no matter how dynamic the impact of SL9 had been, Jupiter was quite at home with energetic processes. But a few days after the last impact, the lights of the southern aurora went out. Gas and dust, rich in complex chemicals and fired into the jovian heights, had drifted on the poleward winds. Faced with an influx of new molecules, our guide did what it always does – made chemistry. So now we shall look at just how H_3^+ was found on Jupiter and elsewhere in the Chemical Cosmos.

Interlude: Trawling for Our Cosmic Guide

Half a world away from Hawaii is the Institut d'Astrophysique de Paris (IAP). It is partly housed in the historic Observatoire de Paris, founded in 1667 and home to such great astronomers as Giovanni Cassini, who demonstrated that what Galileo had described as "ears" on either side of Saturn were a system of rings unparalleled amongst the planets. Among the astronomers currently working at the IAP is Jean-Pierre Maillard, one of the foremost designers of astronomical instruments.

Maillard's instruments have produced stunning results for several decades including the detection of Methane on Mars, a potential indicator of biological activity. In particular, Maillard is world renowned for developing instruments based on "Fourier transform spectroscopy" for visible and infrared astronomy. FTS instruments make use of the wave-like (rather than the particle-like) properties of light, with peaks at the greatest amplitude of the wave, and troughs at the lowest amplitude.

In the FTS technique a beam of light from a star or planet is split into two by a mirror, and the two beams are sent on different paths through the spectrometer. At the point that the beam is split in two, the two beams are said to be "in phase" with one another – the peaks of each beam occur simultaneously. One of the light paths can be varied so that one beam has further to travel than the other. When the two beams are recombined, they are no longer in phase and produce a spectrum of very high resolution like the sound of beats on a musical instrument as strings that are not quite in tune with one another are played together.

One of Maillard's best FTS instruments was set up on the Canada France Hawaii Telescope (CHFT). On the night of September 23, 1988, Pierre Drossart, (now Director of the LESIA laboratory at the Observatory of Paris) Maillard and their colleagues set out to drive the ten miles from the dormitory at Hale Pohaku, which – at a mere 3,000 m altitude – provides a convenient daytime resting place for astronomers in preparation for their nightly observations, to CHFT on the summit of Maunakea. Their target for the night was Jupiter, but not just any part of Jupiter. They were looking for the planet's aurora, a powerful light-show that signifies an enormous and energetic interaction between the planet's magnetic field and interplanetary space.

Aurorae are the stuff of legends. On Earth, shimmering lights of green, red and sometimes blue can be seen at high latitudes in the night sky. Norse and Inuit legends had the aurorae as torchbearers lighting the path of the dead to heaven. They are certainly both beautiful and inspirational. On occasions, when the Earth is hit by a solar storm, aurorae can even be seen in London or Vancouver.

The Aurora Borealis and Aurora Australis – the Northern and Southern Lights – result from the interaction between electrically charged particles from the Sun and the magnetic field that is generated in the center of Earth. Earth's magnetic field – left to its own devices – would be a lot like that of the simple bar magnet, with its north magnetic pole close to the geographical South Pole, and vice versa. Magnetic field lines would cross the Equator parallel to the surface of the planet, skimming the top of the atmosphere. At the (magnetic) poles, the field lines tip almost vertically into the atmosphere.

But the Sun sends out into interplanetary space a continuous, though variable, stream of particles, known as the solar wind. It is not very dense, just a few million atoms for every cubic meter, but it is traveling very fast; at Earth, the solar wind speed ranges between 400 and 750 km/s. Along with it, the solar wind carries a magnetic field. Like an ocean liner ploughing through the waves, Earth with its own magnetic field then "ploughs" through the solar wind. At a distance of about 60,000 km from Earth, a bow shock is formed, and then a boundary that marks the sunward edge of the magnetosphere, that region of space that is controlled by Earth's magnetic field. On the side of Earth facing away from the Sun, the nightside, the magnetic field is distorted in a different way. Instead of being compressed, as it is on the dayside, the nightside magnetic field is dragged by the solar wind away from Earth to form a tail of magnetic field lines stretching hundreds of thousands of kilometers away into space, in the direction of the outer planets. Again like an ocean liner, Earth's passage through the solar wind leaves a magnetic wake behind it.

As a result, Earth's magnetosphere has the shape of a teardrop – or even a comet with its tail – compressed on the side facing the Sun, the dayside, and very much extended on the opposite direction on the nightside. This structure then channels the solar wind particles that flow from the Sun, sweeping them up like a giant vacuum cleaner and firing them along the magnetic field lines. At the poles, as the field lines dive into Earth's atmosphere, the particles follow them down. These highly accelerated, high-energy particles then crash into the atoms and molecules that make up the atmosphere.

In the upper reaches of Earth's atmosphere, lighter gases rise to the top and heavier gases sink down, and there are small amounts of Hydrogen and Helium, reminders of earlier times. The most prominent gas at the top of the atmosphere, however, is atomic Oxygen. As particles crash into this Oxygen, the electrons in the atom jump from state to state, creating beautiful green and red auroral glows. Lower down in the atmosphere the main component is molecular Nitrogen, plus Oxygen, in neutral and positively charged forms. N_2^+ cations can lend a blue tinge to the aurora, if the particles coming in from the magnetosphere are very energetic.

In 1977, NASA launched the twin spacecraft Voyager 1 and Voyager 2 on a mission to explore the worlds of the giant planets: Jupiter, Saturn, Uranus, and Neptune. Thirty years on, and the Voyagers are still sending back vital information about regions of space where the influence of the Sun wanes. Indeed, at the end of 2010, Voyager 1 announced that it had left the Solar System behind; it has now entered interstellar space, where the bubble of plasma given off by the Sun gives way to the currents of the "Local Bubble," created by our neighboring stars past and present.

The year following the launch of the Voyagers, NASA, together with the United Kingdom and the European Space Agency, put the International Ultraviolet Explorer (IUE) satellite into orbit around Earth. IUE's mission was to look at the universe in wavelengths normally screened out by Earth's protective ozone layer. Ten years on, and IUE's string of successes earned its originator, Bob Wilson of University College London, the Medal for Design Excellence from President Ronald Reagan, and a knighthood from Her Majesty's government. Almost simultaneously,

Voyager 1 and IUE reported that Jupiter, like our own planet, had aurorae.

And how! The jovian aurorae were between 100 and 1,000 times brighter than Earth's. Estimates of the energy required to power what the two satellites saw ranged from 10 to 100 terawatts. (A terawatt is 1 million million watts, or a million megawatts.) So Jupiter's aurorae required between 10 million and 100 million megawatts of energy. As a comparison, the Three Gorges power station on China's Yangzte River produces about 20,000 MW. The aurorae that IUE and Voyager had discovered were being powered by a source of energy that would require between 500 and 5,000 Three Gorges power stations.

Everything about Jupiter is big; Jupiter is over 200 times more massive than our Earth. It also comes with an internal magnet that is about 20,000 times stronger than Earth's. That creates an enormous region of space controlled by Jupiter's magnetic field. On the side of the planet facing the Sun, the jovian magnetosphere extends about one million kilometers to its boundary with the solar wind. Downstream from the Sun, on the night side, the tail of the magnetosphere – the magnetotail – goes on for 750 million kilometers, all the way out to the orbit of Saturn. If there was a way of lighting up Jupiter's magnetosphere in the night sky, it would appear 2½ times larger than the full Moon. It is the largest structure in the Solar System, with the exception of the solar wind itself.

The jovian magnetosphere is filled with electrically charged particles, some of them captured from the solar wind, and then trapped until they either crash into the top of the atmosphere or escape down the magnetotail into the outer reaches of the Solar System. Jupiter is a very hostile place. The radiation levels generated by the plasma in the magnetosphere are intense, and spacecraft that venture there have to be especially "hardened" to withstand the jovian environment for any length of time.

Before the Voyagers, two smaller spacecraft, Pioneer 10 and 11 had flown by Jupiter in the winters of 1973 and 1974, respectively. The Voyager 1 spacecraft passed through Jupiter's magnetosphere in the spring of 1979, followed by Voyager 2 in the summer. Neither stayed too close for too long, but went on to Saturn; there is some evidence that when Voyager 2 reached Saturn it was actually inside Jupiter's magnetotail (it is not so easy to escape Jupiter's far-reaching clutches). While, like Earth, Jupiter's magnetosphere does trap particles from the solar wind, that is not its main source of plasma. For many scientists, the greatest surprise of the Voyager missions was not the planet itself, but its moon, Io.

Io is the closest of the four Galilean moons, named for Galileo who discovered them using his newly invented telescope in 1609. Io is slightly larger than our own Moon, and orbits at a daringly close 350,000 km above the surface of the giant planet, closer even than the Moon is to Earth. The enormous gravitational force of Jupiter at such a close distance keeps the insides of Io stirred up like bouillabaisse in a cooking pot. As a result, Io is the most volcanic body we know. Its violent volcanoes spew about one ton of sulfurous gas and dust into space every second, and it is this prodigious volcanism that feeds Jupiter's magnetosphere with plasma. Most of the highly energized particles that eventually crash into Jupiter around the magnetic poles have their origin in Io's volcanoes, not the solar wind.

The aurorae that IUE and the Voyagers saw on Jupiter were not the visible aurorae we see on Earth, but in the ultraviolet. As energized magnetospheric particles plough into the Hydrogen-rich upper levels of Jupiter's atmosphere, they cause the electrons in the atoms and molecules to jump from state to state. As they relax once more, they give off photons of light. Although some visible light is emitted, the resulting emission is 1,000 times stronger at ultraviolet wavelengths. IUE and Voyager were able to show that the emission was coming from around Jupiter's poles, so it was definitely auroral.

Back, then, to that September night on Maunakea, and the Canada France Hawaii Telescope. As well as emitting at ultraviolet wavelengths, due to changes in the state of its electrons, molecular Hydrogen also emits very weakly in the infrared. Maillard and his colleague Pierre Drossart, from the Paris Observatory in Meudon, had reasoned, however, that there was just so much Hydrogen on Jupiter that it ought to be possible for his very sensitive Fourier Transform Spectrometer to detect it – given a long enough exposure time. The plan, therefore, was to point the 4-meter CFHT towards Jupiter's poles and to expose for an hour or so.

The plan worked. At a wavelength of 2.122 μm there was a strong emission line that precisely matched the expected molecular Hydrogen infrared emission. The problem was that it seemed the plan had worked too well, for alongside the molecular Hydrogen spectrum was another series of lines, some of them even stronger than the H_2 lines. Either Hydrogen was *not* the most abundant chemical on Jupiter – which was patent nonsense – or there was a very strong emitter also present high up in the atmosphere around the poles.

An international hunt for the culprit was on, and Takeshi Oka (now at the University of Chicago) and Jim Watson (still at the Herzberg Institute in Ottawa) were to play a key part. Maillard knew that the Herzberg Institute was one of the key places for molecular spectroscopy, and he hoped that Jim Watson might be able to help him identify the mystery lines. Watson was able to tell Maillard's team that they had a spectrum at the Herzberg Institute that looked very similar to his Jupiter auroral spectrum but that they, too, had not been able to identify it. He also contacted Oka; Oka suggested a call to University College London might be in order.

At UCL, Tennyson and his colleagues had perfected a technique for calculating the infrared spectra of small molecules such as Water and H_3^+. Tennyson's team had been working with Oka, identifying the spectrum of our guide, and had recently suggested that new levels of vibrational excitation should provide candidates for spectral lines that could be measured. They proposed that "hot band" lines produced when H_3^+ changed its state by falling from vibrational level 2 to vibrational level 1 would be found amongst the lines formed from the fundamental, level 1 to 0, spectrum. Maillard's Jupiter aurora spectrum – and that in Jim Watson's drawer at the Herzberg Institute – was neither the fundamental nor the hot band of H_3^+, however. It was in the wrong wavelength range, at 2 μm rather than around 4 μm. It was clear, though, that if you added the frequency of the hot band lines to the frequency of the fundamental lines then you did get some of the lines; this was hot H_3^+, high up in Jupiter's auroral atmosphere.

That meant that Tennyson's group had to do a lot more calculating. They had only calculated the lines involving fairly low rotational states. Now they had to calculate the high rotational states that were showing up in Maillard's spectrum, and they had to communicate their results as fast as possible. Today we take the internet for granted; nearly everyone has an email account. Many of us have our own personal, or at least corporate, web pages. Huge work-related documents and personalized Christmas or birthday cards, complete with high-resolution pictures of our friends and families drop regularly into our electronic mailboxes. We post to YouTube, we Skype or iChat, we tweet and blog.

In the 1980s, however, the then "internet" was not very "inter" and consisted of several not very compatible local or national networks. Researchers in the USA had been on Arpanet since the late 1960s, but had only recently switch to using today's TCP/IP protocols for exchanging information. The UK had the Joint Academic Network, JANET. If you wanted to send messages between the two networks, you had to remember that the astronomer's "Starlink" machine at University College London had the address with the "big end" – the country – first, and gradually got more local: uk.ac.ucl.star. The University of Texas astronomers were astro.utexas.edu, with the "small" – local – end first. Other countries were even further behind; a three-line email to Germany could take a week to get through, if it ever did that is.

At the time that Watson contacted Tennyson's group about the Jupiter spectrum Maillard had sent them, in the hope of getting it identified, Tennyson himself was on research leave in Tel Aviv. The rest of the group had been left to mind the store, at least as far as H_3^+ spectroscopy was concerned. Israel was 2 hours ahead of the UK, Ottawa 5 hours behind. A single email message took about 1–2 hours in each direction, so an email exchange took between 2 hours and 4 hours. To get an email from Ottawa via London and on to Tel Aviv, and then get a reply was 4–8 hours.

So Watson got in to work very early in the morning, and sent a list of lines and wavelengths that needed identifying to UCL. The UCL team, by now in the middle of their afternoon, made some preliminary calculations to identify the lines and sent them to Tennyson, working late in his Tel Aviv office. Some refinements to the calculations, and back came the message from Israel for a final check with the UCL calculations (by this time London was well past happy hour), and back went the identifications to

Branching Out: In the Land of the Giants and Dwarves   145

the Herzberg Institute. Two frantic, memorable and enormously exciting weeks went by before all the lines in Maillard's spectrum were properly associated with the quantum labels spectroscopists require before they believe any identification – whether it comes from the Herzberg Institute or not.

Maillard's spectrum was not a normal spectrum in which a molecule goes down (or up, if the light is being absorbed) by just one vibrational bridge at a time. Instead H_3^+ molecules were showing off and jumping two bridges at once, relaxing from vibrational level 2 all the way down to vibrational level 0, without stopping off at vibrational level 1. That was why the frequencies of Maillard's spectrum matched the frequencies of the fundamental band, vibrational level 1 to level 0, *plus* the frequencies of the hot band from level 2 down to level 1.

A spectrum in which the molecule jumps 2 vibrational levels at once is called an overtone. An overtone spectrum has (approximately) twice the frequency, and half the wavelength, of the fundamental, just like a guitar can be made sound its overtone by lightly touching the vibrating string at the midpoint. Overtone spectra are typical of a hot gas; it turned out that the H_3^+ temperature was over 1,000K, very hot for a planetary atmosphere, and adding fuel to a fire that had been burning since the days of the Voyager mission. For Voyager had found that all of the giant planets had temperatures much hotter than expected in their upper atmospheres, a problem that still baffles planetary astronomers to this day.

In the spring of 1992, the United Kingdom Infrared Telescope (UKIRT) was involved in a major survey of H_3^+ emission on Jupiter. The team was led by Larry Trafton from the University of Texas. Trafton was actually the first person to measure the H_3^+ spectrum of Jupiter, a year before Maillard, but his lines were at low resolution and he could not identify them. Nonetheless, he had maintained his interest in the subject, despite his disappointment at being pipped to the post. The observations of jovian H_3^+ Trafton was making went well, but a couple of hours before dawn, Jupiter set below the horizon. That left a dilemma; go back to the Hale Pohaku dormitory early and get more sleep, but leave an unoccupied telescope and be way too early for breakfast (the best meal at Hale Pohaku), or stay and find something else to look at. Trafton suggested looking at Uranus, which was following Jupiter across the night sky.

Given that our adventurer had not yet put in an appearance on Saturn, much closer and bigger than Uranus, and lashed by a much stronger solar wind, it was a long shot. But sometimes long shots pay off, and this one certainly did. The Uranian H_3^+ spectrum showed up clearly, weaker as one might expect, but even more clean and clear than that from Jupiter and showing high temperatures again at nearly 800 K. The next night, Trafton's team tried for Neptune, an even longer shot. This one did not pay off, however, and to date not even the giant Keck telescope has been able to find our chemical guide on the Solar System's outermost planet.

Inevitably, Saturn was next. Later in 1992, Tom Geballe, then Director of UKIRT, used his discretion to check the ringed planet. The H_3^+ spectrum of Saturn was disappointingly weak, just a few percent of that of Jupiter, and surprising given the similarities between the two planets. It completed a "set" for UKIRT, though, and proved the consummate worth of its new spectrometer. (At the time of writing, the British research council that runs UKIRT decided that the telescope will not be supported after 2012, a very sad end to a fine tradition and a decision that we hope will not be irrevocable.)

Although the discovery that H_3^+ was an important molecule in three out of our Solar System's four giant planets was a surprise at the time, it should not have been. For more than a decade, scientists at the University of Michigan's Atmospheric Physics Laboratory, one of the best-known groups in the world, had been using computer models that predicted just this. Their calculations showed that our guide would be copiously produced both by the action of ultraviolet sunlight and by energetic particles as they hit these Hydrogen-rich atmospheres. They were right!

It turns out that our guide plays a schizophrenic role in the upper atmospheres of Jupiter, Saturn and Uranus. Since it is electrically charged, H_3^+ allows currents to flow through the atmosphere, currents – like those running through an electric fire – that *heat* the air. For Jupiter, this heating is equivalent to hundreds of billions of household fires, more than a hundred times greater than the energy provided to the giant planet by ultraviolet sunlight.

If distributed across the whole globe, it could well explain why the upper atmospheres of the giant planets are so hot. But H_3^+ also cools, radiating much of the energy that arrives in the form of high-energy particles and sunlight back out into space – a crucial property that we will look at in the next chapter.

The discovery of H_3^+ emission from Jupiter certainly encouraged those looking for it in its natural home of the Interstellar Medium, but it did not initiate the search. In the 1960s, as the new radio maps of the heavens picked out chemically rich cold dark clouds and hot star-forming nurseries, McDaniel and his Atlanta colleagues had advised astronomers to start looking for H_3^+ out in interstellar space; but what should they be looking for exactly?

Since H_3^+ is a perfectly symmetric equilateral triangle, it does not have the permanent dipole required to produce a radio or microwave signature. If, however, one of the Hydrogen atoms is replaced by its isotope Deuterium to form H_2D^+, the molecule becomes microwave active. Deuterium is twice as heavy as Hydrogen because it has both a proton and a neutron in its nucleus. That means the center of mass of H_2D^+ is shifted away from the center of the triangle towards the heavier Deuterium atom. Since Deuterium is electrically identical to Hydrogen, with just one proton in its nucleus, the positive charge remains at the center of the triangle. The perfect equilateral triangular symmetry is now gone, however, and the small separation between the center of the molecule's mass and its electric charge is enough to produce the dipole necessary for radio astronomy. Maybe the way to track down H_3^+ was through its heavy cousin, H_2D^+.

Although there are only two Deuterium atoms for every 100,000 Hydrogen atoms, the Zero-Point-Energy effect means that in the coldest regions of the ISM, Deuterium-containing molecules are favored. In some clouds, Deuterium-bearing molecules may make up as much as 10% of the molecules found, a concentration of some 5,000. The most convenient spectral line with which to fingerprint H_2D^+ is at the very high-frequency end of the radio spectrum, at just over 372 GHz. This frequency is best observed from a high altitude, preferably by plane or satellite. In January 1985, Tom Phillips and his colleagues from the California Institute of Technology were given a few nights' time on NASA's Kuiper Airborne Observatory.

The very cold Taurus Molecular Cloud was an obvious target for observation, but that turned out to be a no show. They did, however, get an interesting signal from the Christmas Tree Cluster in the faint constellation of Monoceros the Unicorn. This cluster, which bears a passing resemblance to the eponymous fir tree and is known to astronomers by its New Galactic Catalogue (NGC) number of 2264, is a collection of about 40 stars whose starlight bathes the nearby interstellar clouds, causing them to shine pink in the night sky. The pink glow, due to atomic Hydrogen, shows that the outer regions of the clouds are hot. There are, however, some dense clouds where the temperatures are as little as 55K, cool enough for the Zero-Point-Energy effect to make H_2D^+ a suitable candidate for detection. Phillips' results were intriguing; a line at the right frequency and with Doppler shift that was just about right for what was known about the way the gas in the clouds was flowing under the pressure of heating by the nearby stars. Astronomers scratched their heads and said "maybe," waiting still for the real thing, the detection of H_3^+ in interstellar space.

Meanwhile, Oka had set off to find our chemical guide in interstellar clouds just as soon as he had measured the infrared spectrum by which it could be identified. In the summer of 1981, he was on the move from the Herzberg Institute in Ottawa to set up his Ion Factory in Chicago. That slowed him up, but only a little. Also on the move was Tom Geballe, then not long graduated from Berkeley. Geballe was moving to Hilo, Hawaii to become staff astronomer on the United Kingdom Infrared Telescope. Oka sent Geballe a preprint of a paper he had prepared for the UK's Royal Society on the astronomical search for H_3^+. It was the start of a collaboration that has lasted three decades and would take at least half of this time to reach its ultimate goal.

The next year saw the Oka-Geballe team applying for time to use NASA's InfraRed Telescope Facility on Maunakea. If successful, the application promised to provide "the third hydrogenic probe for astronomy after H and H_2". Area 1 in the Orion Molecular Cloud was amongst the targets selected for observation. Unfortunately, although the team were awarded time on the telescope for the night of December 6, 1982, "neither the infrared spectrometer nor the weather was on our side," Oka explained to IRTF Director Eric Becklin. "I believe that H_3^+ is the most important molecular

Branching Out: In the Land of the Giants and Dwarves 149

FIGURE 6.6 An artist's *impression* of the highly energetic young star W33A, with its dark dusty accretion disk and powerful jets of gas pouring out of the top and bottom: *credit – Gemini Observatory, artwork by Lynette Cook.*

species yet to be discovered in space and I will continue my effort on this project," Oka promised.

And so he did. For 15 years, our guide played hide-and-seek throughout the ISM. Weather, spectrometers and telescope allocation panels all seemed to be against Oka and Geballe at one time or another. Finally, on the night of April 29, 1996, tell-tale fingerprints of H_3^+ were found in not one, but two clouds of cold, dense interstellar gas.

The astronomical objects known as W33A (Figure 6.6) and GL2136 are highly energetic and very young stars in the southern hemisphere of our galaxy. They are still surrounded by vast amounts of the gas and dust clouds out which they are forming, and have been studied extensively and mapped in emission from Carbon Monoxide. They have the advantage that some of the compounds of heavier elements that can destroy H_3^+ are probably still frozen out onto the dust grains of the proto-stellar nebula. The two stars are both intense sources of infrared radiation, and the clouds are still cold, an ideal setup for detecting our guide. At high spectral resolution, the spectrometer on UKIRT was able to pick up two key absorptions from H_3^+. It was there at a density of just two molecules for every billion of diatomic molecular Hydrogen, H_2, just 30 molecules in every cubic meter of gas. The lines told Oka and Geballe not only that our guide was present, but that the gas in which it resided had a cold temperature of just 35K. One part of the credo of interstellar chemistry had been proven; H_3^+ is a key and active player in *dense* interstellar clouds. *Nature* published the results, and the UK *Guardian* newspaper reported that "the elusive triangle" had finally been cornered. The story reverberated across the international media.

What was more surprising was the discovery 2 years later that our guide's spectrum could also be found in *diffuse* interstellar gas. These clouds have more or less equal mixtures of atomic and molecular Hydrogen, and have relatively high concentrations of negatively charged electrons that seek out and destroy H_3^+ very rapidly. The astrochemical wisdom was that our guide would be formed in the diffuse ISM, but not as copiously as in dense clouds; it would assist in the chemical production line, but would not hang around long enough to get itself noticed.

Nonetheless, during the summer of 1997, Oka, Geballe and their team had UKIRT and the Mayall Telescope at the Kitt Peak National Observatory pointed towards a very active young star called Cygnus OB2-12, in the constellation of the Swan. They found that our guide was there, clearly identified by the fingerprints that had finally been picked up in W33A and GL2136. Moreover, there was no doubt that the H_3^+ absorption was coming from *diffuse* – not dense – clouds; Cygnus OB2-12 was known to be heavily obscured by clouds but the lack of a tell-tale Water spectrum meant that the gas could not be as dense as in W33A or GL2136. The team also made measurements of the very center of our Milky Way galaxy, towards a region known as the Central Molecular Zone, home to a massive black hole and the most active star-forming region in the galaxy. Recently, this area has been found to be home to two organic molecules of a complexity previously unknown in space: Ethyl Formate and Propyl Cyanide. Once more, there was no doubt at all that our guide was at work in the diffuse and dense clouds towards the Galactic Center, too.

The models that astronomers had worked with for decades to predict how chemistry would work in this crucial part of the ISM had to be wrong; something had to be changed.

One of the numbers many astronomers thought that they had off pat was ten times in a billion billion per second. That was the rate at which cosmic rays, high-energy particles given off by stars and supernovae, would ionize an atom or a molecule in the ISM. Multiply that number by the number density of atoms or molecules (of Hydrogen) in the gas, and it would tell how long it took to ionize one atom or molecule per cubic meter. For a cubic meter of the dense ISM, containing about ten billion molecules of Hydrogen, it would take ten million seconds for one ionization to occur in the gas. For the diffuse ISM with only 100 million per cubic meter, a billion seconds were needed, or nearly 32 years. Although there had been challenges in the past, many still thought the Cosmic Ray Ionization Rate a reliable constant.

The large amounts of H_3^+ found in diffuse clouds by Oka and Geballe meant it could not be squared with such models, however. Either the center of the galaxy had to be nearly 50,000 light years across, equal to the entire radius of the Milky Way – an absurd proposition – or the rate at which cosmic rays ionize the diffuse ISM clearly had to be higher, and its chemistry consequently much more vibrant. How much higher (6-fold, 30-fold or more) is still being debated, but our guide has told astronomers – loud and clear, once and for all – that the Cosmic Ray Ionization Rate is not a constant. Our understanding of the Chemical Cosmos will never be the same again.

7. In the Delta: Exoplanets – Worlds, but Not as We Know Them

Astronomy is about waiting, as much as anything else. We cannot go out and touch the stars, we cannot experiment with them; we just have to take what they send us – light, electromagnetic radiation of all colors and wavelengths. Astronomy is about collecting light, whatever comes our way. Recognizing this implies a passive approach that gives astronomers plenty of time to dream and think about what it is that is doing the shining. Those who are following the adventures of our chemical guide often wonder just where H_3^+ is out there, and what role it might be playing.

One of the recurrent contemplations is whether or not there are other places where life might develop, or might have developed. Science fiction writers clearly have great fun with this – what it might look like and how it might behave, were such life-forms to develop intelligence. There are locations in our own Solar System, outside of Earth, where it is possible that there is, might have been, or might one day be, life. Of the planets, Venus, which is about the same size as Earth, and Mars, which has a radius about half that of Earth's, are obvious candidates at first glance. Venus, however, has a very dense, inhospitable atmosphere of the greenhouse gas Carbon Dioxide, with small amounts of Nitrogen and Sulfuric Acid clouds. The temperature at the surface of Venus is a scorching 740 degrees above absolute zero, more than 450°C. That almost certainly rules out life on Venus, and indeed the planet is often held up as a dire warning of what happens if you have a runaway greenhouse effect when too much Carbon Dioxide builds up.

Mars, on the other hand, has a very thin atmosphere. Again, the main component is Carbon Dioxide, but this time there is so little that the temperature is generally sub-zero. At the poles and during the night, Carbon Dioxide freezes out to form dry ice frosts. There may be water near the surface, and the geology of the planet shows evidence of water having flowed across the Martian surface may be as recently as a few hundred million years ago when giant reptiles and amphibians were colonizing Earth. Finding evidence of life on Mars – existing or fossil – is the goal of several NASA and European Space Agency missions. The next generation may actually see astronauts visiting the Red Planet, on the grounds that the human eye is the most sensitive of all detectors, and that the human brain (still) beats any computer in recognizing patterns and faint clues.

Elsewhere in our Solar System, Jupiter's moon Europa is thought to have an ocean of water slush beneath its icy surface. There just might be enough energy for life to develop. Its neighboring moon, Ganymede, too, has a sub-surface ocean. Titan, the moon of Saturn whose dense atmosphere is rich in hydrocarbons like Methane and Ethane, is another candidate. Titan's atmosphere is often likened to that of the early Earth in "deep freeze". Titan may have to wait a few billion years until our Sun starts to become a Red Giant star before it gets warm enough for some form of life to develop, and even then it may not last long; as the atmosphere warms up, it will start to escape, robbing Titan of the very chemical soup from which life might start to evolve. There may be other niche environments, too, where some sort of life might just get a start, but how far it would then develop is another matter. Hence there are prospects for finding life elsewhere in our Solar System, but they are rather limited. Twenty years ago, astronomers asked if there was life anywhere else but Earth would have been very cautious about saying "yes".

Today, they are still very cautious. With the exception of those who believe that the detection of Methane in the atmosphere of Mars demonstrates that biological activity must have gone on in the not too distant past, even if it is not continuing today, there is no evidence that life outside of Earth exists. But there is a feeling abroad that makes astronomers much more likely to *believe* that there is extra-terrestrial life. The reason for that is the discovery of extra-solar planets, planets that orbit stars other than our Sun. We have now reached the delta of the river of cosmic chemistry, where it branches not just into one Solar System, but myriads, branching over and over again.

Exoplanets, as they are called, are one of the hottest topics in astronomy. They bring together planetary scientists and astrophysicists in an intoxicating mix of great hope and fierce competition. Using the largest telescopes, it is now even possible to take images of exoplanets themselves over one thousand, million, million kilometers away – at least images of what they looked like 130 years ago when the light from them first set out to cross interstellar space. Exoplanets are everywhere we look, and since there are a hundred billion stars in our galaxy alone, even if only a very small percentage of them have planets, there must be billions of planets. Surely, then, Earth cannot be the only planet on which life has evolved.

Isaac Newton speculated that the stars we see in the night sky might be "the centers of other like systems", systems like our Solar System back in the early eighteenth century. Throughout the nineteenth century there were claims that a planet or planets orbited the star known as 70 Ophiuchi, a faint star in the constellation of Ophiuchus the Snakeholder. 70 Ophiuchi is a binary star system, in which two stars, somewhat smaller and cooler than our Sun, orbit around each other once every 88 years. Putting a planet in orbit around such a system was always going to be problematic, and calculations soon showed this system would be very unstable. Nonetheless, 70 Ophiuchi remains a target for exoplanet hunters today. The first definite detection of a planet around another star had to wait until the 1980s, however. Once more, the Maunakea Observatory in Hawaii held the key to finding it; the Canada France Hawaii Telescope scooped the honors, and the Herzberg Institute played its part.

Over hundreds of years, astronomers have tracked the stars in our galaxy, noting their positions with ever increasing accuracy and the rate at which they appear to move in the night sky (or not, as the case may be). Vast catalogues of stars and their astronomical properties have been put together, and nowadays it is bread and butter for researchers to go to the electronic Simbad star catalogue, run from the University of Strasbourg in France, to check on the brightness and positions of any star they want to study, as well as its relative motion with respect to Earth.

If we were able to go deep into space to, say, one of the nearby stars in the constellation of Centaurus, named after the mythical creature that roamed the Labyrinth of Crete, and we looked back at our Sun, we would see it drift very slowly in the heavens. If we had very accurate instruments, however, we might notice that it also appeared to be moving very slightly backwards and forwards along our line of sight. We might see it moving towards us, and then 5 years 11 months later it would be moving away. In another 5 years 11 months time, it would be coming towards us again. The Sun would appear to wobble backwards and forwards with a period of a bit over 11 years 10 months; its relative motion against the stars in the background would change by about 13 m/s in the Centauran sky. The Sun would be responding to the gravitational pull of Jupiter as the giant planet orbited about it, and scientists on an observatory around Proxima Centauri would be able to deduce that our Sun had at least one planet in orbit around it.

Back on Earth, the Herzberg Institute's Bruce Campbell and colleagues had spent many of their nights throughout the 1980s patiently monitoring a number stars with the Canada France Hawaii Telescope, measuring their velocity in the line of sight towards and away from Earth. They had perfected a technique of measuring lines in the spectrum of each star against a very accurate standard, a small cell containing Hydrogen Fluoride gas. Despite its very unpleasant nature – Hydrogen Fluoride is so reactive it is used to etch glass – the gas had been recommended to Campbell by none other than Gerhard Herzberg himself, since it had spectral lines of very well known wavelengths conveniently spread apart so that lots of lines due to the stars themselves could get through. Light coming from the star would be Doppler-shifted to the bluer end of the spectrum, if the star were moving towards Earth and red-shifted if the star were moving away. By comparing star lines with the fixed lines of Hydrogen Fluoride, Campbell's measurement technique was so sensitive it could pick up Doppler shifts due to wobbles of just 10 m/s, an accuracy of 1 part in 30 million.

By 1987, Campbell's team had accumulated enough observations to draw some conclusions: "Seven stars show small, but statistically significant, long-term trends in the relative velocities," they reported in the *Astrophysical Journal*. "These cannot be due to … brown dwarfs … companions of ~1–9 Jupiter masses are inferred … these low-mass objects could represent the tip of the planetary mass spectrum," they announced cautiously.

Campbell's announcement reveals much of the thinking that has to go into making claims about the detection of planets around other stars. Many star systems are binaries, and often a smaller star orbits around a larger one. Brown dwarfs are at the very small end of the spectrum of what counts as a star. They have masses between 13 and 80 times that of Jupiter, which means their centers are too cool to fuse Hydrogen into Helium and burn heavy Hydrogen (Deuterium) instead. Brown dwarfs are, as the name suggests, rather small, dim objects compared with Hydrogen-burning stars, making them difficult to pick out. The gravitational attraction of an unseen brown dwarf orbiting a larger, brighter star would certainly cause the starlight to be Doppler shifted as it was pulled backwards and forwards in the night sky. Nonetheless, Campbell was able to rule them out as the companions to the stars he was monitoring because they were large enough to have already been detected by conventional astronomical techniques. The planet of which the team seemed most sure was around a star called Gamma Cephei, in the constellation of Cepheus, the mythological husband of Cassiopeia and father of Andromeda, whose constellations are also close by in the night sky. This planet, Campbell deduced, had a mass only 1.7 times that of Jupiter, making it very planetary and not at all brown-dwarfish.

Sometimes an idea has to wait for its time. Back in 1988, when Campbell's team reported Jupiter-sized companions for the stars they were looking at, the world of astronomy was not quite ready to believe them.

Millisecond pulsars are at the other end of the family of stars from brown dwarfs. They are neutron stars, remnants of a gigantic supernova explosion that has blasted an enormous star to pieces. Neutron stars are called so because they are made of neutrons and have the density of particles in an atomic nucleus. So, although a neutron star is up to twice as massive as our Sun, its radius is a tiny 12 km, about the size of a city like London and 60,000 times less than the radius of the Sun. This gives neutron stars a density that is about 300 million, million times that of the Sun. As Millisecond Pulsars, these Neutron Stars are spinning on their axis thousands, tens of thousands, of times per minute. They do so with the accuracy of an atomic clock.

FIGURE 7.1 The Arecibo Radio Observatory in Puerto Rico: *credit – NAIC – Arecibo Observatory, a facility of the NSF.*

Since they are the result of a gigantic stellar explosion, Millisecond Pulsars would not appear, at first glance, to be a very hospitable environment for planets. For starters, planets take a while to form in the relatively gentle environment of the Sun (around 100 million years for Earth, although Jupiter and Saturn formed much faster) and the massive stars that give rise to Pulsars live short lives during which they drive away the gas and dust that might go into forming planets as fast as they can. Secondly, any planets that do form are likely to be blasted deep into space when the parent star goes supernova. Nonetheless, the first planet that astronomers felt was believable was found around one of these inhospitable Pulsars.

The Arecibo Observatory in Puerto Rico is home to the largest single dish radio telescope in the world (Figure 7.1). Spanning 305 m, or more memorably 1,000 ft, it fits into a depression left by a natural sinkhole bored into the limestone rocks. Arecibo's iconic radio dish is the stuff of movies, from *James Bond* to *Contact*, and is part of the SETI (Search for Extra-Terrestrial Intelligence) project. Millisecond Pulsars send out powerful beams or radio waves as they spin; perfect targets for Arecibo.

Sometimes astronomers are creative in naming objects, and sometimes they are not. Characters and creatures from mythologies ranging from the ancient Greeks to the North American Indians, via the Norse sagas, have been pressed into the service of identifying whatever shines from the heavens. So it was definitely a bad day in the naming department when the Millisecond Pulsar PSR B1257+12 was christened. When Polish astronomer Aleksander Wolszcznan found it in 1990, however, maybe he did not envisage it taking on the status it now has in the world of exoplanets.

By the end of 1991, Wolszcznan and his colleague Dale Frail, from the National Radio Astronomy Observatory in New Mexico, had completed their analysis of over 4,000 observations of the PSR B1257+12 radio signal. The Pulsar was rotating with a period of 6.21853193177 millisecond (ms), and the uncertainty was 1 in the last decimal place. (If only all watches were that accurate.) There were slight variations, though, in the regularity with which the radio signals were received, of the order of a couple of hundred billionths of a second. These deviations could only be because the Pulsar was being pulled back and forth in the sky by some companion object, or objects. The wobble in PSR B1257+12's position in the heavens was tiny, just about half a centimeter; nonetheless, Wolszcznan and Frail were able to work out that there had to be at least two planets orbiting the Pulsar. Their calculations showed that one planet orbited once every 98 days, and had a mass just under three times that of Earth, while the other, slightly bigger planet went round PSR 1257+12 in just under 67 days.

The mystery was where had these planets come from and how had they formed. Perhaps they were the rocky cores left of what had once been giant planets like Jupiter? Maybe they had formed up anew from a disk of debris left over from the supernova explosion that had destroyed the parent star of PSR B1257+12? But the two astronomers felt moved to speculate: "The detection of a planetary system around a nearby, old neutron star raises the tantalizing possibility that a non-negligible fraction of neutron stars observable as radio pulsars may be orbited by planet-like bodies." Exoplanets had arrived.

That said, exoplanets had arrived where they were least expected, and – in some senses – least wanted. The environment close to a neutron star such as PSR B1257+12 could hardly be said to be the most inviting place. Neutron stars do not shine, their planets would not be bathed in warm sunlight and their powerful radio beams are hardy conducive to fostering the kind of planets where life might develop. What astronomers really wanted was to find a planet or two around a star much more like our Sun, warm and friendly, long-lived and benign. They did not have too long to wait. In October 1995, Swiss astronomers Michel Mayor and Didier Queloz announced that there was a planet orbiting a Sun-like star called 51 Pegasi, nearly 475 million, million kilometers from Earth in the constellation of the flying horse Pegasus.

Although 51 Pegasi is older than our Sun, and a little more massive, Mayor and Queloz had delivered what astronomers had wanted – a planet that was in an environment similar to our own Solar System. Except that it was not very similar. To be sure it had a mass not very different from that of Jupiter, but it was in an orbit that no Solar System planet inhabits today. Writing in the premier science journal *Nature*, Mayor and Queloz remarked: "The [Jupiter mass] companion lies only about eight million kilometers from the star, which would be well inside the orbit of Mercury in our Solar System." That was something of an understatement. At closest approach, Mercury is a toasty 46 million kilometers from the Sun, more than 3 times closer than Earth. 51 Pegasi b, as the planet was prosaically named, was a searing 20 times closer to its parent star than Earth is to the Sun. No wonder that it had a temperature of 1,300K, compared with Earth's mild 300K, 23°C.

Mayor and Queloz had to be cautious – others had claimed exoplanets before only to have to retract their claims. A year before Wolszcznan and Frail had announced their discovery of a planetary system around PSR B1257+12, Andrew Lyne and his colleagues thought they had similarly detected a pulsar planet system, only to have to retract their claim when they realized they had made a mistake in their analysis. Therefore Mayor and Queloz had to take great pains to ensure that they were not making a similar error. Stars are known to have "quakes" like the earthquakes that shake our planet, and they also have spots that can confuse measurements as they rotate into view. Mayor and Queloz did everything they could to rule them out and were still left with a wobble in the starlight from 51 Pegasi. So they were confident in taking on a still skeptical astronomical world. Nonetheless, it was comforting to have the discovery of 51 Pegasi b, as the planet is known, confirmed by Geoff Marcy and colleagues at the Lick Observatory. That still left them, however, the task of explaining the planet.

Our Solar System has small, rocky planets – Mercury, Venus, Earth and Mars – closer in to the Sun, and gas and ice giants – Jupiter and Saturn, Uranus and Neptune – further out. The giants have very large gas atmospheres; the rocky planets much less atmosphere, or none, in the case of Mercury. The giants form when a sufficiently large core of ice (mainly Water ice) forms. (In the case of Jupiter, this core is more than ten times the mass of Earth.) This then grabs the gas left over from the formation of the Sun by gravitational attraction. It is a race against time; at the birth of the Solar System, the hot young Sun is trying to drive the original gas cloud from which it formed out into deep space. As a result, a planet like Jupiter has only a few hundred thousand, maybe a million, years to form its ice core before the gas is all blown away. Therefore large amounts of ice are needed, and you have to be far away enough from the Sun for the temperature to be low enough for Water to freeze. That puts you roughly at the orbit of Jupiter, a little over five times as far away from the Sun as Earth. So it makes sense that the Solar System's gas giants are more distant from the Sun than the rocky planets, which are made up mainly of materials that freeze at much higher temperatures than Water ice.

Mayor and Queloz were faced with the problem that there was no way enough ice could freeze out to form a Jupiter-sized planet a mere 8 million kilometers from a sun-like star. One possibility they considered was that their planet, 51 Pegasi b, was the end product of a brown dwarf companion star that had drifted much too close to 51 Pegasi for comfort and had been stripped of most of its mass. After all, with a temperature of 1,300°, massive evaporation was a distinct possibility. Their other theory was that the planet had, during the course of its history, migrated in from an orbit similar to our Jupiter's to one that was 100 closer than when it had been formed.

This latter interpretation enthused astrophysicists who were trying to create computer models of how planetary systems like ours form. In their simulations, there is still a lot of gas left over from the original cloud even after Jupiter forms up. In many of their models, that gas has enough frictional drag on to cause the planet to drift slowly in towards the Sun. Indeed, it turns out that it is quite difficult to set the conditions just right so that the model "Jupiter" stays where it forms, some 780 million kilometers from the Sun, rather than drifting in to a much closer orbit. Moreover, in many simulations, as "Jupiter" came in from the cold it had a disconcerting tendency to kick its little relatives, like Mars, Earth and Venus, out of the Solar System altogether. So it might be that our Solar System – with Jupiter far away from the Sun and Mars, Earth, Venus, and Mercury safe inside its orbit – is something of a lucky exception rather than the norm. Whatever the case, the discovery of 51 Pegasi b properly ushered in a new era of exoplanetology.

Soon every observatory in the world was working on projects to detect exoplanets. The harvest has been remarkable; from three (or maybe four) in October 1995, the number of confirmed detections had risen more than a hundredfold to around 700 by the end of September 2011, according to the *Extrasolar Planets Encylcopaedia* run by Jean Schneider at the Paris Observatory. The radial velocity method used by Mayor and Queloz, and Campbell before them, along with its close relative, astronometry – which involves noting very slight changes in the position of stars in the heavens – have detected or confirmed over 90% of those planets, with some 80 planetary systems in which there are more than one planet. In addition to this and the pulsar-timing method of Wolszcznan and Frail, however, there are other planet detection methods that are proving productive.

When Albert Einstein formulated his General Theory of Relativity in 1915, one of the predictions that he made was that starlight would be noticeably bent when it passed a very massive object. His prediction was demonstrated in May 1919, during an eclipse of the Sun. The British astronomer Arthur Eddington led a team to make observations in Brazil and on the island of Principe in the

Indian Ocean, from where a total eclipse would be visible. As it happened, stars in the constellation of Taurus the Bull would lie very close to the Sun on the sky. In normal daylight, of course, they would not be visible, but for the brief moments of total eclipse, when the Sun's rays were blotted out by the Moon, they would. Eddington worked out that it ought to be possible to measure the apparent position of these stars and figure out to what extent light coming from them had been bent as it passed our massive Sun. When Eddington, on November 6, 1919, announced that his results confirmed Einstein's predictions, *The Times* ran headlines of a "Revolution in Science", and questions were asked in the British parliament as to whether Newton had been "overthrown".

This light bending effect of large objects comes into its own for exoplanet detection in a technique called gravitational *microlensing*. Such lensing enhances the intensity of light affected by it. At certain times of the year, Earth's orbit will place it such that light from a distant star passes close by one that is much nearer. In those circumstances, the faraway star will have its light "lensed" by the nearer one and will appear brighter. If the lensing star has a planet in orbit around it, then there is a chance that the planet itself will add to the lensing effect, causing microlensing, and there will be a short period during which the faraway star will appear to brighten even more. These effects do not last very long, as Earth's motion around the Sun, as well as the microlensing planet's orbit around its star, mean the microlensing alignment is short lived. A round-the-clock watch is required, but this is a technique that works well with global networks of small telescopes that can be run automatically, feeding their results into centers around the world dedicated to looking out for microlensing events. Not all astronomy needs super large telescopes.

More recently, it has proved possible to detect exoplanets that happen to pass in front of (and behind) their parent star, in the line of sight to Earth. These planets are known as transiting planets. From the standpoint of us here on Earth, Venus and Mercury are transiting planets. That is to say, every so often they pass between us and the Sun, blocking out some of the sunlight we would otherwise receive. Venus crossed the face of the Sun in 2004. (There will be another transit on June 6, 2012, but then we will have to wait until 2117 and 2125 for the next pair.) The transit lasted about 6 h, and the Sun was dimmed by some 0.1%, one part in a thousand. This is not a lot, but certainly enough to be measurable. Transits by extrasolar planets can now be detected routinely. That is the method being used by NASA's Kepler satellite, and it is clear that the floodgates to detection by this method are well and truly open; early in 2011, NASA announced the detection of more than 1,200 new exoplanet candidates, with the promise of thousands more to come.

In 2008, Michel Mayor announced that he had found a planetary system consisting of three "super-Earth's", with masses between four and nine times that of Earth, around a star called HD 40307, a star that is a bit smaller and a bit cooler than our Sun. In 2009, Mayor's team announced a planet with a mass that might be as little as twice that of Earth, and, as 2011 rolled around, some 30 planets with masses between two and ten Earth masses had been found. Kepler's first-announced detections included tens of Earth-size planets, and similar numbers in what is called the Goldilocks, or Habitable, Zone, where conditions are potentially favorable to the evolution of living organisms. One complex system, known as Kepler 11, has six planets orbiting their central star, all with masses in the range between that of Earth and Uranus. Another planet, known as Kepler 16b, has been found orbiting a binary star system – worlds with two "Suns" are not just the stuff of sci-fi movies.

Current detection methods, however, still favor finding large planets, some several times the mass of Jupiter, that orbit close to their central star. These planets have short orbital periods of just a few days meaning that astronomers can observe several orbits in a relatively short period of time. 51 Pegasi b orbits in just over 4 days, for example, compared with nearly 12 years for our Jupiter. Moreover, large planets orbiting close to their central star cause the largest wobbles in the star's radial velocity, towards and away from Earth, which also makes it easier to detect them. If the planet happens to transit across the face of the star, then Jupiters, and super-Jupiters, have a much bigger effect than Venuses or Mercuries; therefore the collection of planets discovered is still somewhat biased towards large planets orbiting close to their star.

That said, the world of exoplanet astronomy is entering yet another new phase, moving from simply detecting planets and saying how big they are to characterizing them in terms of the physics and chemistry going on in their atmospheres. In this field, transiting planets are leading the way, and providing many startling results.

NASA's Spitzer Space Telescope is, in many respects, the forerunner of the James Webb Space Telescope. Launched in the summer of 2003, it has a mirror that is only 85 cm across, giving it just 2% of the JWST's proposed collecting power. This has not stopped it from making many important discoveries during the years of perfect service it has given astronomers, and with the excitement surrounding exoplanets Spitzer clearly had to take a look. One of the planets it chose to observe goes by the unprepossessing name of HD 189733 b. This planet orbits its parent star, HD 189733, some 63 light years away from Earth in the constellation of Vulpecula the Fox, noteworthy only for lying between two bright astronomical birds – Cygnus the Swan and Aquila the Eagle.

HD 189733 b crosses in front of its star, orbiting once every 2 days 5 h 14 min and 15 s (to the nearest second); it orbits its star nearly 165 times every Earth year. HD189733 b skims just 4.6 million kilometers above the surface of its star, more than 32 times closer than Earth is to the Sun. Even though the parent star is cooler than the Sun, the temperature of HD 189733 b's atmosphere is 1,100K or so, some 800K hotter than Earth This planet is about 13% bigger than Jupiter, and must have a very similar composition to our Solar System's giant.

Spitzer can carry out measurements making use of several filters that allow infrared light of various wavelengths through. If astronomers could fit the resulting spectrum to possible models of the planet's atmosphere, they could work out what else, apart from Hydrogen and Helium, it contained. Working at the Institut d'Astrophysique de Paris, Giovanna Tinetti was given the job of trying to make sense of what Sptizer provided. It was going nowhere. But at the annual conference of the newly formed Europlanet network of European planetary scientists, she met up with one of Jonathan Tennyson's group, who had just carried out calculations of what the spectrum of Water would look like at high temperatures, like those in the atmosphere of HD 189733 b. Including the high-temperature Water data produced a near-perfect match to the planet's spectrum; *Nature* published the result in July 2007.

It was the first detection of a molecule in a planet outside of our Solar System. It was Water, the key chemical for life. Tinetti's team followed up their results with observations by the NASA Hubble Space Telescope. Not only did Hubble confirm the Water detection, it showed that HD 189733 b also contained Methane, another key life-endowing molecule, and it may be that Ammonia is also present. That said, no one is claiming that life – at least as we conceive of it – could exist on such a searingly hot Jupiter-like planet. These key detections show us, though, that the Chemical Cosmos works on exoplanets as it does those in our own Solar System.

Given that so many of the exoplanet astronomers are now trying to characterize are much like Jupiter (huge spheres of Hydrogen gas), it is fair to ask what role our chemical guide, H_3^+, might play in terms of their chemical composition and overall behavior.

HD 209458 b in the constellation of Pegasus was the very first exoplanet to be detected transiting across the face of its parent star. It is one that has provoked enough interest for astronomers to give it a name – Osiris, the Egyptian god of the afterlife. This planet has a mass about 70% that of Jupiter, but its radius is about 30% greater, which gives it a density more like Saturn. Like HD 189733 b, Osiris is very close to its parent star, orbiting just 7 million kilometers from the center once every 3½ days. In 2001, David Charbonneau, of Harvard University, reported that there was Sodium vapor in the planet's atmosphere. This was followed up by Alfred Vidal-Madjar, from the Institut d'Astrophysique in Paris, who found that it also had a much extended atmosphere, stretching out to 2½ times the planet's official radius. Moreover, Osiris was losing atmosphere fast; over 100,000 tons of Hydrogen was escaping into space every second. Indeed, there were some calculations showing that Osiris might evaporate altogether, although more sober reflection shows that it has lost probably about 7% of its mass during its 5-billion-year lifetime.

Planets face a daily battle to keep hold of their atmospheres. It is a battle between the effects of the Sun, or their star, to boil off the atmosphere, and the force of their gravity to hang onto it. In our Solar System, Mars, which is some nine times less massive than Earth, has pretty much lost that battle. Earth has lost, and continues to lose, its light gases like Hydrogen. Jupiter, with over 300 times the mass of Earth, does a pretty good job of keeping everything on board. As well as the mass of the planet, and hence its gravity, how much sunlight it receives makes a big difference. The hotter the atmosphere the more likely it is to escape; in extreme cases, the atmosphere may expand so fast that it exceeds the escape velocity and, like a space-bound rocket, it literally takes off.

Jupiter orbits the Sun about 780 million kilometers distant. Osiris is orbiting at less than 1% of this distance. Since the amount of sunlight or starlight received by a planet goes up rapidly, at a rate proportional to the distance squared, as you approach a star, this means that Osiris is being blasted by roughly 10,000 times as much light as Jupiter. It is going to get very hot. A third factor in the equation, however, is how efficiently a planet is able to cool itself down, either by reflecting the starlight or by re-radiating it, and that depends on the chemical composition of its atmosphere, among other things. For atmospheres mainly made of Hydrogen, that cooling process is a problem.

Diatomic Hydrogen molecules, H_2, are very poor radiators since they vibrate they do not change their symmetry as they vibrate. Hence their vibrations are simply not able to generate much infrared radiation to get rid of the heat building up in a planet's atmosphere. Hydrogen atoms only start to be useful when the temperature gets up to nearly 25,000K, by which time Jupiter would have expanded to three times its present radius, a bit larger than the extended atmosphere of Osiris. Our chemical guide, H_3^+, on the other hand, is a very good radiator. Additionally, more sunlight, or starlight, means more ionization of diatomic Hydrogen molecules, which, in turn, generates more H_3^+, meaning more cooling. This "H_3^+ thermostat" has been found to work for the giant planets in our Solar System and ought to be transferable to other planetary systems.

At UCL, the Atmospheric Physics group led by Alan Aylward thought they could test this. There were two clear observations. Jupiter, orbiting the Sun far out in the Solar System, had a stable, atmosphere that only extended out to about one tenth of the planet's nominal radius. Osiris, on the other hand, orbiting close to its star, had an unstable, highly expanded atmosphere. Suppose you brought Jupiter in towards the Sun – at what point would it start to look like Osiris, and at what point would its atmosphere go unstable, too? And how would this happen, would it be a gradual change, or would it happen all of a sudden?

Aylward's computer models of Jupiter's atmosphere run in three dimensions. That is vitally important because winds are very good at taking energy away from hot spots, where it is building up with potentially destructive consequences, to colder regions, where it would just have the gently warming effect of a breeze from the Tropics. The models also allow for the chemistry of the atmosphere to change. That is also vitally important. For starters, the amount of H_3^+ produced would depend on the amount of sunlight received. Secondly, if the temperature did start to build up, then the main atmosphere of Jupiter would begin to change from diatomic Hydrogen molecules to single Hydrogen atoms. Tommi Koskinen, the student running the computer experiment, found he had to make several other refinements to Aylward's models so they would remain stable under the extreme conditions he was about to subject them to.

When he finally got them, Koskinen's results were dramatic. He discovered that you could bring Jupiter five times closer to the Sun than it is now, equivalent to the Earth's orbit, and the temperature in the planet's upper atmosphere remained a balmy 1,200K or so. Infrared cooling by the now-much-augmented H_3^+ cations matched whatever the Sun could throw at Jupiter by way of ultraviolet heating. Bring it in another five times closer, and the H_3^+ thermostat continued to do its stuff. Jupiter was now at a mere 30 million kilometers from the Sun, nearly twice as close as Mercury is today. Each million kilometers closer was becoming that bit more painful, though. At 24 million kilometers from the Sun, the thermostat still worked, but only just so; the temperature was now up to 3,000K.

Just past 23 million kilometers the game was up. A chain reaction started – not quite enough H_3^+ could form to keep the planet cool, and as the temperature started to rise diatomic Hydrogen molecules shook themselves to pieces. And as they broke up into individual Hydrogen atoms, the essential fuel for making new H_3^+ was lost. It was a vicious cycle of breaking Hydrogen molecules and non-forming H_3^+, with the temperature soaring rapidly. At 21 million kilometers from the Sun, the temperature had skyrocketed to over 20,000K, and the planet's atmosphere had expanded to over twice its original size. Osiris-like conditions were starting; Koskinen's model, based on assumptions of stability, had had enough.

What Koskinen had shown, however, was important to understanding the behavior of Jupiter-like exoplanets, the so-called hot Jupiters. If they migrated in towards their central star, they might stop before this sharp boundary – this stability limit between Jupiter-like or Osiris-like, was reached, or they might carry on through it. Just when they would encounter it would depend on many factors, not least of which was how much ultraviolet radiation their central star would be throwing in their direction. That might also depend on when, exactly, they started to spiral in, if they did. For when stars evolve, they do not always shine with the same brightness, particularly when it comes to ultraviolet radiation.

The "Sun in Time" project has attempted to work out how our Sun's radiation might have changed as it evolved from a bright young thing to the more sedate, middle-aged star that it is now. This project has made careful measurements of a number of Sun-like stars whose ages can be reliably estimated. Their results show that the infant Sun, aged just 100 million years and not the 4.6 billion it is now, would have given out over 100 times more ultraviolet radiation than it does today. That would mean the stability limit for Jupiter, before its atmosphere "blew up" into an Osiris-like planet, would have been not 23 million kilometers, but 250 million kilometers, well past the orbit of Earth and a third of the way out to Jupiter's present position.

The "Sun in Time" project also predicts that every time you reduce the Sun's age by a factor of ten, its ultraviolet intensity goes up by a factor of sixteen. If that trend were to continue back to the Sun's baby years, say just 15 million years after it formed, Jupiter would be close to the stability limit, with its atmosphere on the point of escaping into space again. Could cooling by our chemical guide, our H_3^+ thermostat, have saved Jupiter's bacon in those early years of the Solar System? If so, then our chemical guide has played a critical role in shaping our Solar System as we see it today. For without Jupiter, life may never even have got going here on Earth.

8. Towards the Sea of Life

This book set out with our guide H_3^+ to take us on a journey from the early universe, the source of the river of cosmic chemistry, down the rapids of exploding giant stars, through the muddy depths of the Interstellar Medium, to the formation of planets and moons, and, with them, the potential for life to come into existence. This is a journey through the Chemical Cosmos – it sets the scene for everything living to come into being, and the laws that govern the Chemical Cosmos still hold for the living universe, too. Approaching the mouth of the river, where it flows into the sea of life, we have to say farewell to our guide, however. From now on, H_3^+ watches our progress from the lighthouses of giant planet atmospheres, or from the oceans of interstellar space.

Living things are much more than their chemical components and their reactions. With life, qualities change with quantity and complexity in ways that make all attempts to reduce them to "just a bag of chemical reactions" fruitless and absurd. Although biological complexity cannot escape the more fundamental laws of chemistry, living things use them in ways that non-living matter cannot. Nonetheless, all living matter requires non-living matter for its existence. Plants absorb Carbon Dioxide from which they produce complex sugars and starches; most animals require Oxygen to burn their food and generate energy; and as far as we know at present, all living things need water in one form or another. The inorganic becomes organic and vital.

The honor of demonstrating this is usually attributed to the German chemist Friedrich Wöhler who, in 1828, synthesized Urea, the waste product mammals excrete in urine. Wöhler had started from Cyanic Acid and Ammonia, both *inorganic* chemicals. At the time, Wöhler was working in Stockholm with the famous Swedish chemist, Jöns Jacob Berzelius. Berzelius was one of those who thought that it was too simplistic to look on living organisms as if they were just some kind of machine and that there had to be some *vis vitae*, or some vital force at work. Wöhler's synthesis of Urea demonstrated that even if there were, the gap between the inorganic world and that of organisms, the organic, was not unbridgeable. As it has developed, organic chemistry has produced synthetic medicines, dyes, and even structural materials – but not yet life.

Earth is the only place in the universe that we can be sure life exists; whatever our beliefs may be, we have found it nowhere else. Earth is the "Goldilocks planet" – conditions here really are "just right". Earth is not, like Venus, too hot, or, like Mars, too cold. In the 1920s the Russian scientist, Alexander Oparin, and the British biologist, J.B.S. Haldane, independently suggested that life could arise under certain conditions if there were "pools" of organically rich, pre-biotic, material on the surface of the early Earth. Those organic chemicals would have to be produced from whatever was terrestrially available. The Oxygen-rich air that Earth has today is mainly the product of biological activity. In its early years, Earth's atmosphere was rather different, probably dominated by Nitrogen and Carbon Dioxide.

In the 1950s, Harold Urey, by then at the University of Chicago, thought that the early Earth's atmosphere was probably made mainly of Methane and Ammonia. But, like today's "best guess", he thought it unlikely any free Oxygen was present. Urey speculated that pre-biotic molecules might be made direct from the air if enough energy were supplied. Chicago Ph.D. student, Stanley Miller, persuaded Urey to allow him to carry out a series of experiments to see just what could be produced from such an atmosphere, in the presence of Water and Hydrogen, using electrical discharges to simulate the effect of lightning. From his first experiments in 1953, Miller was able to positively identify the Amino Acids Glycine and Alanine, key components of proteins, amongst the products of the discharge tube. There were other biologically important molecules, too, but their traces were too faint to be absolutely certain. The press, picking up on the story, speculated that Miller would soon be producing beef-steaks.

By the time that Miller and Urey reviewed their own work and that of other like-minded scientists in 1959, they could list twenty biologically active compounds that had been identified in such electrical discharge experiments, including Wöhler's Urea.

The two chemists also pointed to experiments that showed solid surfaces, such as clays, might help simple Amino Acids join together to form even more complicated molecules, such as Peptides. These surfaces might act as a local venue for a life-forming mix of pre-biotics to collect, before taking the final – quite literally – vital step. Their paper in *Science* went further, to discuss the possibilities of life on Venus (probably too hot) and Mars (probably too cold). That said, they concluded: "All the projected space flights and the high costs of such developments would be fully justified if they were able to establish the existence of life on either Mars or Venus. In that case, the thesis that life develops spontaneously when the conditions are favorable would be far more firmly established, and our whole view of the origin of life would be confirmed."

Subsequent missions and ground-based observations have pretty much ruled out Venus – it really is too hot, and its atmosphere just too nasty. Prebiotic organic molecules must have been fried to a crisp or vaporized into little pieces well before they could get anywhere along the path to life. Mars, on the other hand, is intriguing astronomers the world over. Yes, it is mainly too cold: although temperatures get fairly warm in sunlight, they dip down to below the freezing point of Carbon Dioxide, some 80K below the freezing point of Water, in the dark and at the winter poles. For centuries, however, there has been speculation that Mars could host life. H.G. Wells famously made use of this in his 1898 *War of the Worlds* novel; Orson Welles notoriously terrified Americans with his Halloween 1938 radio broadcast of the book.

In 1976, a pair of spaceprobes, Viking 1 and Viking 2, landed on Mars. For over 6 years, scientists had the benefit of data beamed back from Viking 1, although Viking 2 lasted only just over half that time before its batteries failed. The landers carried instruments designed to test for life on Mars. To start with, it looked as if they had found it; the Labeled Release experiment let a drop of water containing nutrients that had been "labeled" with radioactive Carbon-14 onto the martian soil. There appeared to be biological activity; radioactive Carbon Dioxide, $^{14}CO_2$ the expected product of an organism making use of the nutrients, was detected. Other Viking experiments, however, failed to find the organic molecules that should have been on the surface of Mars had life really been there. It is now generally felt that the $^{14}CO_2$ detected was the product of an *in*organic reaction in the martian soil, but the enthusiasm to search for life on the Red Planet has not been diminished.

Christmas Day 2003 should have been one of great celebration in the household of Colin Pillinger, Professor of Planetary Sciences at the Open University. His pride and joy, a lander named Beagle 2 after the ship that carried Charles Darwin on his revolutionary voyage to the Galapagos Isles, should have touched down on the surface of Mars. Beagle 2 had been put together on a shoestring budget, hitched a ride on the European Space Agency's Mars Express, and was due to deliver at least a preliminary verdict on whether the Red Planet was, or was still, biologically active. But space is a very dangerous place. The last that was heard of the lander was that it had separated effectively from Mars Express 5 days before its expected touchdown and was well on its way to the surface. Whether Beagle 2 crash-landed or whether, like a stone skimming across a pond, it "bounced" off the thin martian atmosphere and hurtled away into interplanetary space, will never be known for certain. What is certain, though, is that it never called home.

As a lander, Beagle 2 was a bust. What it achieved, however, was remarkable for future missions. At the heart of the lander was a device known as the Gas Analysis Package, or GAP, a miniature chemical laboratory intended to sort out the various gases in the atmosphere of Mars close to its surface and those that could be released by digging a little way into the ground and boiling them out of the red dirt. At the heart of GAP was a mass spectrometer, similar to the cumbersome apparatus used nearly a century previously to find our chemical guide, but this time only the size of a saucer. Beagle 2 pointed the way in miniaturization from which other missions are benefiting in their search for signs of organic, pre-biotic chemistry on Mars and elsewhere. NASA and ESA, with their ambitious Mars-exploration programs, are building on the Beagle 2 legacy.

There is something of an edict for those looking for signs of life or pre-biotic chemicals: "follow the water". The martian atmosphere is too thin to keep water liquid on the surface of the Red Planet; even if the temperature gets above freezing, it just boils off into space. Water might exist in the martian version of the Siberian permafrost, and there seems near-irrefutable evidence that water flowed on the surface of the Red Planet at some time in the past, though whether that past should be measured in millions or billions of years is a matter for fierce debate. In the summer of 2009, scientists working on NASA's Phoenix lander, deliberately sent to the north pole of Mars, announced that they had definitely detected water in one of their martian soil samples.

Elsewhere in our Solar System, Jupiter's moon Europa is attracting attention. Europa is the smallest of the four Galilean moons, with a diameter of just over 3,100 km, about 360 km less than our own Moon. It orbits just 600,000 km above the surface of the giant planet, in a hostile environment of electrically charged gas – plasma – that is swept past it by Jupiter's powerful magnetic field. Voyager images showed a smooth, but cracked, billiard-ball body with dark materials filling the cracks in an otherwise icy blue surface. Taking a closer look, the Galileo spacecraft found evidence of the moon being geologically active. The surface had far fewer impact craters than the neighboring moons, Ganymede and Callisto, a sign that it had been "resurfaced" by volcanism in the last few tens of millions of years. There appeared to be rafts of Water ice that had cracked and turned like a European version of Arctic pack-ice. Moreover, Galileo instruments measuring the magnetic fields around Jupiter picked up a signature from Europa that changed depending where the moon was with respect to the planet.

Although a completely solid moon just might exhibit all of these characteristics, it is now thought that Europa has an ocean of salt-water slush beneath its icy surface. If it is there, the ocean is probably around 100 km thick and lies about a kilometer beneath the surface. Europa, which formed alongside Jupiter from the proto-solar nebula, must have Carbon-based molecules in its chemical mix. According to Chris Chyba of the SETI Institute, it has probably also received some billion tons of Carbon-bearing molecules over its lifetime, about one thousandth of the biomass of Earth. Ganymede, too, may have such a slush ocean. And there just might be enough energy for life to develop; sometime in the next decade or so, if all goes well, Europe and/or the United States will send a space mission back to have another look at Europa and Ganymede.

Still following the water, a joint NASA-ESA mission has recently found a vast fountain spilling from the icy moon, Enceladus. For while Beagle 2 was crashing and burning, another double act was sliding quietly in to position, not over Mars but much further away, at the giant planet Saturn and its equally impressive moon, Titan. More than 25 years since it was first mooted, 7 years after it was launched and 1½ billion kilometers from Earth, the Cassini-Huygens mission got into position for the most ambitious of all long-distance encounters. Exactly a year after Beagle 2 should have landed on Mars, Huygens (a specially designed descent probe) set off from the main Cassini orbiting spacecraft. On January 14, 2005, it plunged into the atmosphere of Titan on a trip to the surface of a very alien world; unlike Beagle 2, Huygens made it in one piece. Still 120 km up, Huygens was buffeted by winds of over 400 km an hour, racing round the moon in the same direction as Titan was rotating. The atmosphere at this altitude was super-rotating, as computer models and Earth-based observations had predicted.

Alien, Titan might be. Nonetheless as Huygens surveyed the land around it, scientists watching the descent at the European Space Agency's Darmstadt control center were impressed by how Earthlike everything appeared – hills and dunes, "river" channels, even lake-like features complete with shorelines. There were also craters testifying to Titan's bombardment by bits and pieces left over from its giant neighbor, Saturn, and, perhaps, evidence of volcanoes, not of hot magma as on Earth, but of ices similar to those seen on Jupiter's moon Europa – cryo-volcanoes. After a descent lasting 2½ h, Huygens touched down on a surface like "crème brulée"; the wind had dropped to just a few km per hour, the humidity was around 50%. There were pebbles and rocks such as one might find in a dried-up riverbed back home (Figure 8.1).

Titan was discovered by the Dutch astronomer Christiaan Huygens in 1655, and the lander was named for him. With a diameter of just over 5,000 km, Titan is the second largest moon in the Solar System, after Jupiter's Ganymede. Were Titan in orbit around the Sun, and not Saturn, it would qualify for the status of planet, for it is larger than Mercury, although less massive, weighing in at just 2¼% of the mass of Earth. Titan orbits the ringed planet at nearly 1¼ million kilometers, such that sometimes it is inside Saturn's magnetosphere and sometimes just outside, exposed to the

FIGURE 8.1 The Huygen's probe's view of its landing site on Titan: *credit – the European Space Agency*.

full fury of the solar wind. Additionally, Titan has an atmosphere that extends out hundreds of kilometers into space (some 10% of the moon's diameter) and is ten times denser than Earth's, creating a surface pressure 1½ times that at sea level – quite an achievement.

And what an atmosphere it is!

According to the veteran planetary scientist Toby Owen of Hawaii's Institute for Astronomy, Titan is the "Peter Pan" world, potentially a genuinely grown up planet, like Earth, but left in a Never-Never-Land limbo, a case of severely arrested development. Like Earth, Nitrogen dominates the titanic atmosphere. In place of our Oxygen, Titan has large amounts of Methane; the moon's Oxygen content is largely locked up as Water ice on its surface, where the temperature is a shivering 180° below freezing, or in the rocks below. Analyses of the small amounts of the rare, inert gas Argon show that the Nitrogen present is probably due to the breakdown of Ammonia and similar compounds over billions of years, an atmosphere on the way to becoming Earth-like but still lacking the key signature of terrestrial biological activity – free, gaseous Oxygen. Nearly there, but not quite.

How Titan keeps its current atmosphere, and for how long it will keep it, are problems that stretch the ingenuity of planetary scientists to its limit. Methane is the first problem. Prior to the arrival of Cassini/Huygens, the Voyagers had found out the basics of Titan's atmosphere – Nitrogen at the bottom, Hydrogen Cyanide at the top, and Methane everywhere else. Sunlight plays chemical havoc with such a mix, and the Methane is constantly being used as fuel for more complicated organic molecules. Left to its own devices, the amount of Methane currently in Titan's atmosphere would last just 10 to 20 million years, so it would have been used up billions of years ago. The only explanation is that Methane is being put back into the atmosphere more or less as fast as sunlight is turning it into other things. Prior to the arrival of Cassini/Huygens the explanation for this was that the surface of Titan was covered with vast lakes and seas of HydroCarbons composed largely of liquid Methane and Ethane. Liquid reservoirs would replenish the atmosphere, and were themselves being refilled by Methane/Ethane rain from clouds that had been discovered hovering about 15 km above the ground. It's a good story.

On its way down to the surface the Huygens lander had taken pictures of lake-like and river-like features. They looked somewhat dry, though, and the orbiting Cassini spacecraft's radar instruments failed to pick up the sort of signals that wet rivers, wet lakes and wet seas ought to make. Pictures taken in visible and infrared wavelengths were similarly indecisive. Finally after 2 years orbiting around Saturn and after several close encounters with Titan, Cassini did manage to pick up the right signals. In the northern latitudes there were lakes in impact craters and old volcanic calderas like Clearwater Lake in Canada and Crater Lake in Oregon. Those images made Titan only the second place in the universe that we know has *liquid* on its surface; none of the other planets or moons in the Solar System has surface liquids. At the same time, models trying to make sense of the information coming back to Earth from Cassini predicted that there should be a planet-wide Methane drizzle accompanied by storms and flash floods that would carve the river valleys Huygens saw.

So the good story remained a good story except there still did not seem to be sufficient lakes to supply enough Methane to keep the cycle going. Moreover, what clouds there were seemed to be confined to the southern, not northern, hemisphere in two bands, one at latitudes equivalent to the tropics on Earth, another at poleward, temperate latitudes. While waiting for Cassini to find the big lakes, Jonathan Lunine, of Rome's Universitá degli Studi, and his team at the Universities of Arizona and Nantes, had put forward another explanation for the Methane. This gave the moon a highly active geological history that involved Methane being released from the rocks from which Titan formed in three major episodes.

The first started not long after Titan became Titan, rather than a collection of rocks not required on board the good planet Saturn, and lasted some billion years. The original center of Titan, according to this theory, would have rocks containing large amounts of the gas that quickly formed a "clathrate" – an icy mixture of Water and Methane – light enough to cause the core to turn over like an overloaded freight tanker, dumping heavy Silicate rocks at the center. This released the pressure and allowed large amounts of Methane to escape into the moon's atmosphere. Titan had to wait until it was some 2 billion years old for the second episode to come into its own. This required the once-rocky Silicate core to become turbulent enough for heat to escape from the center, warm the clathrate above it and allow more Methane gas to percolate to the surface and into the atmosphere. Finally, the last billion years of Titan's geology if the model is right, have been characterized by plumes of Methane gas escaping through the moon's icy mantle as part of its cryo-volcanic activity.

Volcanism, even icy cryo-volcanism, needs heat, however. Although Titan has lost all of the heat generated when it formed, and used up that which would have been available from the decay of radioactive isotopes, it is close enough that the gravitational pull of the giant planet Saturn generates what is called "tidal heating". That tidal heating does seem to be sufficient to power icy volcanoes that, in turn, could be pumping Methane into the moon's atmosphere. After several years of scrutinizing the images from Cassini,

FIGURE 8.2 NASA has constructed this image of Sotra Facula a cryovolcano on the surface of Titan, based on data returned by the radio science instruments on Cassini: *credit – NASA.*

it does seem that the spacecraft has sent back compelling evidence that such volcanoes really do exist on Titan; a 1,000-m high peak, known as Sotra is the best candidate such a fiery beast (Figure 8.2).

That's another good story, and Lunine's explanation for the current levels of Methane may be at least part of the picture, if not the whole deal. But it is also clear from the latest Cassini images that, at least in the northern hemisphere at high latitudes, there are genuine seas of Methane/Ethane Hydrocarbon mixtures the size of the American Great Lakes. Whether or not the Methane reservoirs (however they are generated) will run out, Titan's current atmosphere is proving to be one of the most exciting places in the Chemical Cosmos. Sunlight breaks down Methane to form Ethane and Acetylene. Adding Nitrogen into the mix produces Hydrogen Cyanide, HCN, and the even nastier Cyanogen, C_2N_2. From these active starting materials, Butane, Poly-Acetylene and Cyano-Acetylene are just a hair's breadth away. More exotic Hydro-Carbons, like C_3H_3, pair up to make Benzene, C_6H_6.

For ground-based creatures like us, the tendency is to look upwards. What happens higher up in the atmosphere is a consequence of what happens below. Hot air rises, Water evaporates and forms clouds as it cools, Oxygen is released by biological activity on land and in the seas and drifts upwards, and other geological and biological processes release Carbon Dioxide from rocks, plants and fossil fuels. True, sunlight does come down to us from above, as does rain. More generally, however, if "it" is going anywhere, what starts low can only go higher. So, generally speaking, there is more of "it" lower down, near the source, than higher up. It has been something of a surprise, therefore, to find that Benzene is evenly distributed throughout Titan's atmosphere; there is as much near the top as there is lower down. Moreover, Titan shows a great tendency for chemicals to form at the top of the atmosphere, which then play an important role in what goes on below.

Before it became compulsory to burn only smokeless fuels in London, the coal fires of the capital of England used to produce chemical smogs ("pea-soupers") that set the scene for many a murky mystery. In the slums of the Victorian city's notorious East End, Jack the Ripper slid undetected past groping policemen, hidden by the gloom. Hawaii's active volcano Kilauea is currently covering the island with a thick, Sulfur-rich, volcanic fog that irritates the throat and makes the scenic views hazy and indistinct. Titan's atmosphere is the foggiest, however, putting these 'earthly' fogs to shame. Beautiful Cassini images show layer after layer of hazes disarmingly reflecting the sunlight in a rainbow of colors that belies the true griminess of the moon's air. Much of that airborne organic grime is produced high in the atmosphere, where electrically charged molecules can dance a chemical fandango of frenetic tempo. Once formed, these heavy, grimy molecules drift leisurely downwards.

Titan's upper atmosphere, some 1,000 km above the surface, is seared by ultraviolet rays from the Sun and bombarded by high energy plasma particles that come either directly from the solar wind or from inside Saturn's own magnetosphere. These ionizing conditions form a series of positive cations, like our chemical guide, but because Titan is too small to have held on to its original Hydrogen, *protonated* H_2 (H_3^+) is not there. Much heavier cations, ones that Titan's gravity can cling onto, predominate. Most abundant is *protonated* Hydrogen Cyanide, $HCNH^+$, together with *protonated* Methane, CH_5^+ and Ethylene, $C_2H_5^+$, and the bizarre

$C_3H_3^+$ species. Altogether fifty or so positive cations containing anything up to eight Carbon atoms have been identified. While our guide has a mass divided by a charge ratio of just 3, the largest positive ion on Titan has a hefty mass/charge ratio of 350. It is this fearsome ionic brew that produces Benzene at high altitudes. And once Benzene is formed, PAHs (Polycyclic Aromatic Hydrocarbons) are an almost foregone conclusion.

Even more unexpected has been the range of *negative* anions discovered by Cassini. At the lighter end of the scale are the Cyanide anion, CN^-, the Amide anion, NH_2^-, and the Benzene anion, $C_6H_5^-$. At the heavy end it gets *really* heavy – some of the negative anions discovered on Titan weigh in with a mass/charge ratio of several thousand, and by the end of 2010 the record stood at mass/charge 13,800! This has led Cassini scientist Jack Hunter Waite to propose that these ultra-heavy ions are charged versions of massive organic molecules known as Tholins. The name Tholin means – among other things – "muddy"; it was proposed by Carl Sagan at the time of the Voyager missions to describe the "brown sticky residue" that he and his colleagues had been producing in their discharge tubes in updated versions of the Miller-Urey experiments. Sagan had predicted that such materials would be fairly common in chemically active astronomical bodies, such as the Interstellar Medium.

Toby Owen's description of Titan as the Peter Pan of the Solar System, an Earth in "deep freeze", begs the question as to whether Titan would make the transition from prebiotic to biotic chemistry were it to be warmed up. This is not just idle speculation; in approximately 4½ billion years time, our Sun will cease to be the benign yellow dwarf star that we know and love and will swell up to become a Red Giant, like the Bull's Eye Aldebaran or the Hoku Lea Arcturus. As it does so, Earth and the other inner planets, Mercury and Venus, will probably be engulfed by its outer layers. Titan, on the other hand, might just warm up enough for the chemistry of life to begin. That, unfortunately, takes us to the second problem for Titan in keeping its atmosphere – at a higher temperature more of the atmosphere will boil away. It will be a race between the evolutionary dynamics of life and evolutionary dynamics of the atmosphere, and the outcome is uncertain.

What studies of Titan do show convincingly, however, is that atmospheric chemistry coupled to a more-or-less continuous source of basic chemicals such as Methane, supplied either from lakes or from volcanic plumes, produces a rich, pre-biotic chemistry. We do not know what is going on in the lakes or on the moon's surface; if, or rather when, we send a mission back to Titan, there is no doubt we shall be surprised. That is the lesson of every space mission sent to the other worlds of the Solar System.

But if Titan were to become life-bearing (assuming it is not already so and we just have not found the evidence) its chemistry would still have to go a long way from where it is today. The original Miller-Urey experiments of the 1950s showed that Amino Acids could be formed from inorganic starting materials that they thought were present in the early Earth's atmosphere of Methane and Nitrogen. Although it is now thought that the ancestral atmosphere was probably Nitrogen and Carbon Dioxide plus Water, similar experiments using the "correct" air mixture continue to produce Amino Acids in significant amounts. So proteins could be formed if these Amino Acids could be persuaded to get together.

Today, that persuasion is supplied by genes composed of the self-replicating molecules RiboNucleic Acid (RNA) and Deoxyribo-Nucleic Acid (DNA). Nowadays RNA and DNA are produced exclusively by biological organisms, and they then go on to produce more biological organisms. Before there was life, however, either RNA, DNA or something rather similar must have been formed chemically, rather than bio-chemically.

From the end of World War II, there was a race to determine the structures of the genetic super-molecules RNA and DNA. The race involved some the world's most eminent scientists, such as Linus Pauling, Rosalind Franklin and Maurice Wilkins. It was won, in 1953, by Cambridge scientists Francis Crick and James Watson. Their Nobel Prize-winning structure consists of a spiral backbone made up of sugar-like molecules linked together through negatively charged Phosphate anions, derived from Phosphoric Acid. The sugar-like molecule, which contains five Carbon atoms, is called Ribose, in RNA, and, with the loss an Oxygen atom, Deoxy-Ribose, in DNA. Here, Carbon chemistry shows the importance of the property known as chirality, or handedness. RNA and DNA make use of the right-handed forms of their backbone-forming sugar to create a right-handed spiral. The Phosphate anions consist of a Phosphorus atom surrounded by four Oxygen atoms. These link to active sites on the (Deoxy-) Ribose molecules on either side of them.

Attached to the Ribose-Phosphate backbone of RNA and DNA are "bases", aromatic rings containing Carbon and Nitrogen atoms. For RNA, these bases are Adenine and Guanine, Cytosine and Uracil; in DNA Uracil is replaced by Thymine. Cytosine, Thymine and Uracil are members of a chemical family called Pyrimidines, based on a single Benzene-like ring of atoms, in which two of the six Carbons have been replaced by Nitrogen. Adenine and Guanine are two of a family of molecules called Purines, which have the Pyrimidine ring of four Carbon atoms and two Nitrogen atoms fused with a smaller Carbon-Nitrogen five atom ring. RNA and DNA form a "double helix", two strands wound round each other, with their backbones on the outside, and the strands held together by chemical bonds between the bases. Carbon chemistry produces highly shaped molecules, and to form self-replicating bodies, the shapes have to be just right. In RNA, Guanine always pairs up with Uracil and Adenine always pairs up with Cytosine. In DNA the pairings are Guanine to Cytosine and Adenine to Thymine. These Carbon-Nitrogen nucleobases have structures that give them specific shapes, and the backbone positions them so that they fit together like keys into locks.

Fairly simple chemistry involving Hydrogen Cyanide, HCN, joining with itself repeatedly gives rise to the Purine Adenine, which contains nothing but Hydrogen, Carbon and Nitrogen. The other key Purine, Guanine, which also has an Oxygen atom hanging off one of its rings, can also be formed by small variations on the HCN theme, with a little Water added into the mix. On the surface of the early Earth, pools of chemically rich seawater may have provided the ideal conditions for such reactions. Cyano-Acetylene, HC_3N, found in the ISM and Titan, is required to make the Pyrimidines Thymine and Uracil; like HCN, HC_3N can join up with itself to form the required rings. Cytosine seems hard to make from essentially pre-biotic mixtures that may have existed on the early Earth. This has given rise to the idea that the final steps towards living things took place in an "RNA world" in which RNA was the information carrying genetic material.

For the backbone, Phosphorus is required in the form of Phosphates. In Earth's crust, Phosphorus is largely tied up as the rock

Fluorapatite, which forms into beautiful hexagonal crystals of colors ranging from green to violet. Only small amounts would have been in solution in the Achaean rock pools – may be enough, maybe not. The hardest part of synthesizing RNA appears to be the formation of Ribose, the sugar that links with the Phosphate anions. While other sugars can be readily made from simple materials like Formaldehyde, Ribose is not among them. In those early years, it is possible that other sugar-like molecules, or even non-sugars, played the role that Ribose plays today.

Bringing enough pre-biotic molecules of the right sort together fast enough, even in chemically concentrated pools, remains a challenge to the laboratory and theoretical chemists who are trying to simulate the pre-biotic conditions under which life stood even a small chance of evolving. Present-day organisms are heavily dependent on enzymes, or biological catalysts, to make its reactions go at the pace of life. One lesson from the ISM is that grain surfaces can speed things up even in the most chemically unpromising conditions. Another is that reactions involving electrically charged ions also go faster, when time is of the essence. On Earth, clay surfaces that can attract electrically charged molecules may have played the role of grains in the ISM. Other candidates for the primitive biochemical laboratory include the "Black Smokers", volcanic vents on the floor of the oceans that have ecosystems that survive even in the complete absence of sunlight, using geothermal energy to power the necessary vital reactions. Therefore pre-biotic molecules could have formed independently from basic inorganic materials here on Earth.

That said, it does seem a shame – from a terrestrial and human point of view – to have the Interstellar Medium go to all the trouble of producing complex organic molecules, to have them neatly wrapped up in comets and meteorites and then delivered to Earth in vast amounts by Jupiter during the Heavy Bombardment period, just for the early Earth to go through all the bother of synthesizing them out of inorganic molecules, be they on the ground or in the atmosphere. Yes, Miller, Urey and all their followers proved that it *could* be done, but was it *necessary*?

In the first 500 million years after Earth formed, it was bombarded so heavily by meteorites and, particularly, comets that they could have delivered Carbon-bearing molecules at the rate of a million tons a year, a staggering 5,000 billion tons over that period, five times Earth's current biomass. Of course, nowhere near all of that would have been in the form of biologically important compounds. And how much of the pre-biotic component would have survived the searing heat of entering Earth's atmosphere is unknown. Nonetheless, the discovery of Uracil in the Murchison meteorite shows that some Nucleobases, at least, can survive the shock of an impact on Earth; laboratory experiments now show that some Amino Acids could also have made it all the way down to the surface intact.

In 2008, a 4-m asteroid known as 2008 TC$_3$ was tracked as it entered Earth's atmosphere, so that the resulting fragments could be swiftly collected from the Nubian Desert in Sudan before contamination could result. The students from the University of Khartoum who made up much of the collection force and gathered nearly ten kilograms of fragments, were amazed to see the diversity of material that 2008 TC3 produced. In that material, the team led by Muawia Shaddad in Khartoum and Peter Jenniskens of the SETI Institute, have found 19 of the 20 Amino Acids used by life on Earth. Moreover, the recent detection of GlycolAldehyde, a simple sugar and precursor of Ribose, in the ISM may help to solve the problem of putting the backbone into RNA and DNA, supposing it could have been delivered intact to Earth.

Just how "organic" could an impactor get? That was *the* question in August 1996 when a team of scientists led by David McKay of NASA's Lyndon B. Johnson Space Center in Houston published their analysis of the meteorite known simply as ALH84001. This meteorite was not primitive, like Allende. It did not date to the origin of Solar System, but to a time some half a billion years later. Nor was it a remnant left over from the formation of the planets, but a chip off Mars, struck off when the Red Planet itself was hit by some marauding body. After a final kick to get it into orbit, it had wandered the Solar System for 15 million years, putting up with the abuses of radiation and, perhaps, other collisions. About 13,000 years ago, ALH84001 crossed paths with Earth and landed in the Alan Hills region of Antarctica. There it was found, the first sample from the region picked up in 1984.

McKay's team had scanned ALH84001 with a powerful electron microscope and had found traces of PAHs. They also found tiny structures, highly reminiscent of bacteria, but on a much smaller scale. McKay and his coworkers came to the conclusion that these were micro-fossils, traces of living creatures that could only have evolved on Mars itself. The Chemical Cosmos had just become a bio-Chemical Cosmos no longer limited to just one planet. Their work was published in the world-leading journal *Science*. It sparked a furious debate; *Science*'s own Richard Kerr warned that "extraordinary claims require extraordinary evidence". Nonetheless, President Bill Clinton was sufficiently inspired to announce a major program of space exploration for Mars, and billions of dollars followed his new enthusiasm.

ALH84001 is both an inspiration and a warning. It raised the possibility that biotic molecules and fossils – even, as a minority of astronomers have claimed, simple bacteria – could survive a journey through space to reach Earth. But the meteorite's 13,000-year sojourn, even in a place as pristine as the Antarctic has still raised questions as to whether samples containing organic or even biotic materials were actually only showing just how easy it is to contaminate meteoritic messengers from space. The debate rumbles on; Mars is still at the forefront of exploration missions, fossils or not.

The more biological aspects of the Chemical Cosmos generate enormous and furious discussions about what is and what is not reasonable. In October 2006, a special school on "complexity" was held in Erice on the Mediterranean island of Sicily. Rome University scientists Pasquale Stano and Pier Luigi Luisi centered the meeting on a number of key questions that – even if currently unanswerable, or unconvincingly answered – would provide foci for further research. These questions give a taste of where research into the boundary between the Chemical Cosmos and the realm of bio-chemistry and life is going.

Foremost among the questions was to what extent laboratory experiments were a good guide to what happens in biotic or pre-biotic systems. In the lab, the chemicals that result from reactions tend to be the most energetically, or *thermodynamically*, stable under equilibrium conditions, when everything has settled down comfortably. Life and the chemistry of living, on the other hand, is far from equilibrium and involves rapid reactions *kinetically* controlled and driven by biological catalysts known as enzymes, so a chemical mix brewed up in the lab is never likely to produce life because it simply lacks these catalysts. The Roman scientists wanted to know what the pre-biotic bridge between thermodynamic and kinetic might look like.

Two more of the questions posed at Erice have long puzzled biochemists. Living organisms on Earth make use of 20 of the family Amino Acids. Yet incalculable numbers of these chemical compounds exist. Why these 20, then – pure chance or was something more fundamental behind the choice made by terrestrial life-forms? Moreover, life on Earth has shown a preference where chirality is concerned, preferring left-handed Amino Acids over their right-handed cousins – again, pure chance or something more fundamental?

Next in Stano and Luisi's firing line was the so-called RNA world. Was it merely a plaything for synthetic biologists, with no relevance to the actual evolution of life? And, anyway, to get self-replicating molecules, like the tango, takes two, they said. The chances of two RNA molecules (assuming they could be formed in the first place) coming together *prior* to some kind of life existing seemed pretty slim. Going for RNA self-replication simply because Uracil was easier to make than Cytosine seemed to fall from the frying pan of pre-biotic chemistry into the fire of biotic chemistry; the problem was merely shifted not solved. Additionally the term "biotic" itself raised issues of just how complex chemistry had to be to be life. After all, even the simplest cells today have 500 genes to provide the information to make the required complement of proteins. Could life be or have been simpler; did theoretical models help solve this riddle? Just how old was the genetic code that we have today, and – a bio-molecular version of the chicken/egg debate – did genes precede cells or cells precede genes?

These questions illustrate how much remains to be learned on the chemical/biochemical interface. Yet the fact that Stano and Luisi *could* ask these questions with the expectation of at least partial answers in the years ahead is testimony to the vitality of this interface and the progress that is being made in understanding it. Scientists have at least made it into the realm of known unknowns, although there are almost certainly some unknown unknowns waiting to trip them up.

Amongst scientists looking at the issues surrounding the production of pre-biotic molecules and them then taking the next step for life to evolve – opinions are very divided as to how easy this "final" step really is. If, as some believe, that step is all but inevitable, one might ask the following question: "Given that Earth is a planet hospitable to life, then did it evolve just once here, or did many 'life's' develop?" Our "life" – Life-1, carbon-based, and centering either on DNA or RNA – is the only one we have found so far on our home planet. But could there be niche habitats where Life-2 (or 3 or 4 ...) clings on despite the predominance of Life-1?

In December 2010, NASA astrobiologist Felisa Wolfe-Simon and her team reported that they had found a strange bacterium, christened GFAJ-1, that could make use of normally poisonous Arsenic, instead of the usual element Phosphorus, in key parts of its biochemistry. The strange bacterium was discovered in California's Lake Mono, which has very high levels of Arsenic in it. Deprived of Phosphorus, it seems that GFAJ-1 can replace the Phosphorus in its energy producing chemical pathway and – critically – in the backbone of its DNA, although how stable this is a question for further research. None of the team is claiming that this is "alien" life or the sought-after Life-2. But it is a sure sign that the river that flows from the Big Bang can follow detours and diversions on its journey to the Sea of Life.

Epilogue

Through the chapters of this book, we have now traced the Chemical Cosmos from its simple source in the Era of Recombination a few hundred years after the Big Bang to its mouth, at the point that it pours its accumulated waters and all their muddy contents into the complex oceans of the *bio*chemical universe. We have dipped our toes into the salty brine. Our guide, H_3^+, has taken us most of the way, showing us how complex chemical pathways are initiated and subtle physical effects are produced by the simplest of molecules – a long journey for a small molecule. H_3^+ has proved itself to be stable enough, to be tough enough, to endure the hardships of interstellar space and the radiation-filled atmospheres of the giant planets in our Solar System and beyond.

It has also, however, shown itself to be highly reactive, quite prepared to sacrifice itself to the greater good of the Chemical Cosmos, usually by handing on a proton to form new ionic species. H_3^+ starts and mediates a web of chemical reactions allowing simple molecules to form and combine all the way to the pre-biotic compounds needed for life to develop. Like all molecules taking part in chemistry, H_3^+ goes through a phase of being both itself and not itself before it finally quits the stage, and makes way for other actors.

Chemistry is about atoms and the combinations of atoms called molecules. It is also about how these combinations break up and recombine; chemistry just would not be chemistry if everything were perfectly stable and stayed the same. What, then, does such a molecule "look" like at the point it is giving up the ghost? If chemists cannot answer that question for a molecule as simple as our guide, can they really be said to understand how any reacting molecule behaves?

In 1982, Southampton University chemist Alan Carrington carried out a, conceptually, remarkably simple experiment. First make your H_3^+. That is easy – pass a strong electric current through a discharge tube containing molecular Hydrogen gas. Then excite the resulting H_3^+ molecule until it has just not quite enough energy to release its proton. Allow the now highly agitated molecule to drift along a tube where it can be hit with an infrared laser beam with enough energy in a single photon to push H_3^+ over the edge so that it breaks up into a diatomic Hydrogen molecule, H_2, and a proton, H^+. Finally, collect the resulting proton and measure its energy.

In practice, the experiment was not quite so simple, and the machine room of the Chemistry Department was put through its paces for many years before Carrington got just the equipment he needed. The infrared laser could be tuned to wavelengths between 9 and 11½ microns in the mid-infrared, not a great range, but enough. The number of protons collected at each wavelength was carefully recorded, producing a spectrum throughout the laser's range. Amazingly Carrington and his team recorded 26,500 individual lines in their spectrum. It was a mess, though, seemingly random and chaotic. When they went to pieces, molecules really did seem to lose it badly.

When Takeshi Oka had measured the first H_3^+ spectrum at the Herzberg Institute in 1980, he had been measuring changes in vibrations at the bottom of the potential energy valley, from Vibrational Bridge 0 to Vibrational Bridge 1. Jumps like that are fairly easy to understand, using simple theories and models. Oka's spectral lines could be assigned with Quantum Numbers, labels that proclaimed just how much vibrational energy was in the molecule, and how much rotational. Carrington, on the other hand, had pushed our guide to the top of the valley, to the region just below the plateau, where the concepts of vibrations used to label the bridges give way to a meaningless jumble as the molecule starts to break up and dissociate.

He was, however, able to say something about the pre-dissociating states from which his laser beam was pushing the molecule to destruction. They had a lifetime of the order of a microsecond, one millionth of a second. Carrington was also able to work out that the final state of the H_3^+ molecule could only last between a few nanoseconds, billionths of a second, to just under a microsecond. To understand Carrington's spectrum required theoretical chemists to produce states that lasted long enough to fit the experiment's requirements, without lasting too long, and to provide them in sufficient numbers. Carrington did provide one additional clue; at low resolution, the 26,500 individual lines collapsed into a few regular, manageable clumps. There might be some order in the chaos, some dignity in death.

At first, the theorists tried states that had very large amounts of rotational energy. These might be produced if the proton, H^+, was flying around the remaining two Hydrogens at a great distance like a ball on a string. If the "string" broke as a result of the laser then the proton would fly off to take part in chemical reactions and a diatomic molecule of Hydrogen, H_2, would be left. Or high-energy rotational states could be produced by the molecule rotating rigidly altogether at high speed. The problem was that not enough high-energy rotational states with the right lifetimes could be made to account for the Carrington spectrum. Nor could the lower resolution clumps be explained.

It fell to Eli Pollack of the Weizmann Institute in Israel to point the way. He showed that highly energetic vibrations, not rotations, might fit the bill. In this scenario, rather than spinning so fast that bits flew off it, H_3^+ was literally shaking a proton free. Pollak teamed up with UCL's Jonathan Tennyson to carry out a proper Quantum Mechanical calculation of the problem. Most of the highly excited vibrational states they produced looked totally chaotic, just a mess. But they found some, in which even though the triangular H_3^+ molecule was bending so energetically, so violently, that it became linear, there remained regular features that could account for Carrington's low-resolution spectrum. Amazingly, even at the point of breaking up, our chemical guide was behaving itself with decorum.

Way to go for such a simple little molecule!

Annotated References and Further Reading to Chapters

1. During the course of researching this book, I made use of literally hundreds of scientific papers and articles, as well as referring to a number of "standard" texts. It is not helpful to list them all here. But in what follows, I list some of the more notable and/or helpful of the material I read. This will give readers keen to follow up on the subjects raised at least a starting point for further reading and research of their own. And I make no bones about it: I use Wikipedia – I use it mainly to check that I really do know what I think I know. Nowadays, many Wikipedia articles are well referenced and enable interested readers to delve further into subjects that interest them.

Chapter 1

1) Much of the background to this chapter, and others, comes from my own experiences and discussions with Professor Takeshi Oka, of the University of Chicago, and my work with Professor Jonathan Tennyson, of University College London. Their impressive lists of publications can be found at:

http://fermi.uchicago.edu/publications/ and http://fermi.uchicago.edu/publications/cv.shtml

http://www.ucl.ac.uk/phys/amopp/people/jonathan_tennyson/papers

2) Another useful reference was:

James F.A.J.L. *Michael Faraday: a very short introduction.* (Oxford University Press, 2010).

Frank James' book does exactly what you would expect from the title, and is a useful background to Michael Faraday's contribution to science and – in Chapter 4 – to electrochemistry and its nomenclature.

Chapter 2

1) Bouwens R.J. and 11 co-authors "A candidate redshift z~10 galaxy and rapid changes in that population at an age of 500 Myr". *Nature* **469**, 504–507 (2011).

The galaxy reported here is the oldest so far discovered, and dates to a period when the first stars were re-ionizing gas in the early universe. Analysis of stars in it show that stellar densities increased rapidly over the next 200 million years.

2) Ciardi B. and Ferrara A. "The first cosmic structures and their effects". *Space Science Reviews* **116**, 625–705 (2005).

Among other issues, this excellent review discusses the mechanisms by which stars form in the early universe and their sizes.

3) Flower D.R. and Pineau de Forets G. "The thermal balance of the first structures in the primordial gas". *Monthly Notices of the Royal Astronomical Society* **323**, 672–676 (2001).

This paper draws attention to the role of cooling by H_2D^+ when primordial gas clouds are shocked as they collapse in on themselves.

4) Gamov G. "Expanding Universe and the Origin of Elements". *Physical Review Letters* **70**, 7–8 (1946).

In this paper, Gamow first points out that conditions for "rapid nuclear reactions" would have existed for only a very short time, explaining why the abundance of elements falls off as the element gets heavier.

5) Glover, S.C.O. and Slavin, D.W. "Is H_3^+ cooling ever important in primordial gas?". *Monthly Notices of the Royal Astronomical Society* **393**, 911–948 (2009).

This paper gives a detailed analysis of the way the primordial gas clouds cool to form the first stars in the early universe. It comes to the conclusion that H_3^+ never accounts for more than 10% of the total cooling required.

6) Lepp S., Stancil P.C. and Dalgarno A. "Atomic and molecular processes in the Early Universe". *Journal of Physics B* **35**, R57–R80 (2002).

This review discusses the mechanisms by which neutral atoms form in the Recombination Era and the formation of the first molecules in the universe, including pathways by which H_3^+ is both created and then destroyed.

7) http://www.astro.ucla.edu/~wright/CosmoCalc.html

This website is an incredibly useful one for turning redshifts into ages after the Big Bang, and *vice versa*. It allows for various assumptions about the nature of the universe.

Chapter 3

1) Alpher R.A., Bethe H. and Gamow G. "The origin of chemical elements". *Physical Review* **73**, 803–804 (1948).

This is the original "alpha, beta, gamma" paper, to which Hans Bethe allegedly contributed only his name, that proposes a "big bang" origin for the chemical elements.

2) Arnett W.D. and Bahcall J.N. "Supernova 1987A". *Annual Reviews of Astronomy and Astrophysics* **27**, 629–700 (1989).

Arnett and Bahcall cover the evolution of Supernova 1987A, the brightest seen from Earth for 383 years, over two years from its first detection. They examine data taken at a large range of wavelengths.

3) Burbidge E.M., Burbidge G.R., Fowler W.A. and Hoyle F. "Synthesis of the elements in stars". *Reviews of Modern Physics* **29**, 548–650.

"BBFH" is the *tour de force* through which Fred Hoyle and his team demonstrated that the chemical elements heavier than

Lithium could be synthesized in stellar nuclei, and introduced the Triple-Alpha process to explain the formation of Carbon.

4) Cherchneff I. and Lilly S. "Molecules in nearby and primordial supernovae", in *Low metallicity star formation: from the first stars to dwarf galaxies; Proceedings of the International Astronomical Union Symposium No. 251*, 221–225 (2008).

Cherchneff and Lilly consider that primordial supernovae may have delivered much of their material to space in the form of molecules, especially CO and SiO.

5) Cox P. "Molecular gas at high redshift" in *Astrochemistry: Recent success and current challenges; Proceedings of the International Astronomical Union Symposium No 231*, 291–300 (2005).

This paper shows that the molecule CO, Carbon Monoxide, was present in galaxies as early as 870 million years after the Big Bang.

6) Glover S.C.O, Clark P.C., Greif T.H., Johnson J.L., Bromm V., Klessen R.S. and Stacy A. "Open questions in the study of population III star formation", in *Low metallicity star formation: from the first stars to dwarf galaxies; Proceedings of the International Astronomical Union Symposium No. 255*, 3–17 (2008).

Population III stars are the oldest in the universe, and – according to this review – most were very massive, with just a few stars much smaller than our Sun.

7) Gribbin J. and Gribbin M. "Stardust: the cosmic recycling of stars, planets and people" (Penguin Press, London, 2000, 2009).

John and Mary Gribbin give a much more detailed account of the discovery of the synthesis of elements in stars in their book than can be given here in just one chapter.

8) Harris G.J., Lynas-Gray A.E., Miller S. and Tennyson J. "The effect of the electron donor H_3^+ on the pre-main sequence and main sequence evolution of low-mass, zero-matellicity stars". *Astrophysical Journal* **600**, 1025–1034 (2004).

For small primordial stars, this paper shows that H_3^+ causes stars with a mass less than 40% of the Sun to evolve more slowly than previously calculated.

9) Meikle W.P.S., Allen D.A., Spiromylio J. and Varani G.-F. "Spectroscopy of Supernova 1987A at 1-5μm I. the first year". *Monthly Notices of the Royal Astronomical Society* **238**, 193–223 (1989).

Among other features, this paper contained the question-marked 3.4 and 3.5 micron infrared lines of SN1987A, later identified as due to H_3^+ ...

10) Miller S., Tennyson J., Lepp S. and Dalgarno A. "Identification of features due to H_3^+ in the infrared spectrum of supernova 1987A. *Nature* **355**, 420–422 (1992).

... and this was the paper that reported the identification of the Meikle lines.

11) Soderberg K.M. and 23 co-authors. " A relativistic type 1bc type supernova without a detected γ-ray burst". *Nature* **463**, 513–515 (2010).

This paper discusses the evolution of a very large, very energetic supernova.

Interlude 1

For this Interlude, a large number of papers could be cited. I have chosen the following arranged in the historical order with which they cover key moments in the discovery of H_3^+.

1) J.J. Thomson's discovery of the H_3^+ molecular ion is covered in three key papers:

1a) Thomson J.J. "Further experiments with positive ions". *Philosophical Magazine* **24**, 209–219 (1912).

1b) Thomson J.J. "Some further applications of the method of positive rays". *Nature* **91**, 333–337 (1913).

1c) Thomson J.J. "Rays of positive electricity". *Proceedings of the Royal Society* **A89**, 1–20 (1913).

2) And they are described by Takeshi Oka in:

2a) Oka T. "The H_3^+ ion" in *Molecular Ions: spectroscopy, structure and chemistry*, in Miller T.A and Bondybey V.E. (eds.), 73–90 (North Holland Publishing Company, 1983).

2b) Oka T. "The infrared spectrum of H_3^+ in laboratory and space plasmas". *Reviews of Modern Physics* **64**, 1141–1149 (1992).

3) Hogness T.R. and Lunn E.G. "The ionization of Hydrogen by electron impact as interpreted by positive ray analysis". *Physical Reviews* **26**, 44–55 (1925).

This is the paper that outlines the key reactions for forming H_3^+ in molecular Hydrogen gas.

4) Joseph Hirschfelder and Henry Eyring set out their key calculations of the energy and structure of H_3^+ and its parent molecule H_3 in a series of papers from 1936–38:

4a) Hirschfelder J., Eyring H. and Rosen N. "I. Calculation of energy of H_3 molecule". *Journal of Chemical Physics* **4**, 121–130 (1936).

4b) Hirschfelder J., Eyring H. and Rosen N. "II. Calculation of energy of H_3^+ ion". *Journal of Chemical Physics* **4**, 130–133 (1936).

4c) Hirschfelder J., Diamond H. and Eyring H. "Calculation of the energy of H_3 and H_3^+. III.". *Journal of Chemical Physics* **5**, 695–703 (1937).

4d) Stevenson D. and Hirschfelder J. "The structure of H_3, H_3^+ and of H_3^-. IV.". *Journal of Chemical Physics* **5**, 695–703 (1937).

4e) Hirschfelder J.O. "The energy of the triatomic Hydrogen molecule and ion, V.". *Journal of Chemical Physics* **6**, 795–806 (1938).

5) A number of workers in the 1960s and 1970s worked on the calculation of the energy and structure of H_3^+, demonstrating that the correct and most stable geometry was that of an equilateral triangle, including (in date order):

5a) Christoffersen R.E., Hagstrom S. and Prosser F. "H_3^+ molecular ion: its structure and energy". *Journal of Chemical Physics* **40**, 236–237 (1964).

Annotated References and Further Reading to Chapters   201

5b) Conroy H. "Potential energy surfaces for the H_3^+ molecular ion". *Journal of Chemical Physics* **40**, 603–604 (1964).

5c) Hoyland J.R. "Two-center wavefunctions for ABH_n systems: illustrative calculations on H_3^+ and H_3". *Journal of Chemical Physics* **41**, 1370–1376 (1964).

5d) Joshi B.D. "Study of the H_3^+ molecule using self-consistent-field one-center expansion approximation". *Journal of Chemical Physics* **44**, 3627–3631 (1966).

5e) Carney G.D. and Porter R.N. "H_3^+: geometry dependence of electronic properties". *Journal of Chemical Physics* **60**, 4251–4264 (1974).

5f) Carney G.D. and Porter R.N. "H_3^+: *ab initio* calculation of the vibration spectrum". *Journal of Chemical Physics* **65**, 3547–3565 (1976).

6) Gaillard M.J. and 9 co-authors "Experimental determination of the structure of H_3^+". *Physical Review* **A17**, 1797–1805 (1978).

This paper reported experiments that confirmed the structure of H_3^+.

7) Oka T. "Observation of the infrared spectrum of H_3^+". *Physical Review Letters* **45**, 531–534 (1980).

This was the paper through which H_3^+ truly announced its presence beyond all doubt to the World. Oka used the 1976 paper by Carney and Porter (above) as his guide, but had to make allowances for the inaccuracies in their calculations.

Chapter 4

1) Three older books are very helpful for providing a scientific background to understanding the Interstellar Medium (ISM), all involving Professor David A. Williams:

1a) Duley W.W. and Williams D.A. *Interstellar chemistry*. (Cambridge University Press, 1984).

1b) Dyson J.E. and Williams D.A. *The physics of the interstellar medium*. (Institute of Physics Publishing, Bristol, 1980).

1c) Hartquist T.W. and Williams D.A. *The chemically controlled cosmos.* (Cambridge University Press, 1995).

2) Much of the information about basic Carbon chemistry – bonding types, isomerism, optical activity *etc.* – can be found in any standard organic chemistry textbook, such as Clayden J., Greeves N., Warren S. and Wothers P. *Organic chemistry* (Oxford University Press, 2001) or Winter A. *Organic chemistry for dummies I* (Wiley Publishing Inc., Indianapolis, 2005).

3) The proceedings of two recent International Astronomical Union symposia – Number 231 *Astrochemistry: Recent success and current challenges* (2005) and Number 251 *Organic matter in space* (2008) – provide lots of information that gives a good indication of the state of astrochemistry at present. So several papers from these symposia are referred to here, although many more helped in the writing of this chapter.

4) Charney S.B. and Rodgers S.D. "Pathways to molecular complexity" in *Astrochemistry: Recent success and current challenges; Proceedings of the International Astronomical Union Symposium No 231*, 237–246 (2005).

This paper outlines series of chemical reactions that can give rise to complex Carbon-bearing molecules, as a result of grains, dust and gas-phase chemistry.

5) Davies R.D., de Jager G. and Vershuur G.L. "Detection of linear and circular polarization in the OH emission sources near W3 and W49". *Nature* **209**, 974–977 (1966).

This paper represents an early success for the astronomers at the Joddrell Bank radio telescope in detecting polarized light from dense molecular clouds.

6) van Dishoek E. "Organic matter in space – an overview" in *Organic matter in space; Proceedings of the International Astronomical Union Symposium No 251*, 357–365 (2008).

Ewine van Dishoek looks at the range of environments in which organic molecules are found in space. Some of the questions she feels are still open include the role of UV radiation in modifying these molecules and what is the main form of solid carbon in space.

Annotated References and Further Reading to Chapters 203

7) Genzel R. and Stutzki J. "The Orion molecular cloud and star-forming region". *Annual Reviews of Astronomy and Astrophysics* **27**, 41–85 (1989).

This review gives a detailed insight into the physics and astrochemistry of the huge OMC; the first protostars were discovered there, along with most of the molecules known in space.

8) Martin D.W., McDaniel E.W. and Meeks M.L. "On the possible occurrence of H_3^+ in interstellar space". *Astrophysical Journal* **134**, 1012–1013 (1961).

These Atlanta-based physicists use their laboratory experiments to suggest that H_3^+ should be present in the interstellar medium.

9) Matthews CN. and Minard R.D. "Hydrogen Cyanide polymers connect cosmochemistry and biochemistry" in *Organic matter in space; Proceedings of the International Astronomical Union Symposium No 251*, 453–457 (2008).

The title of this paper makes clear the importance these authors attach to Hydrogen Cyanide in the formation of prebiotic molecules such as Amino Acids and nucleobases.

10) Millar T.J. "What do we know and what do we need to know" in *Astrochemistry: Recent success and current challenges; Proceedings of the International Astronomical Union Symposium No 231*, 77–86 (2005).

Tom Millar argues that astrochemists still need more sophisticated models of grains and physical conditions in the interstellar medium, points echoed by David Williams, in his summing up of the conference: Williams D.A. "Future directions in Astrochemistry" in *Astrochemistry: Recent success and current challenges; Proceedings of the International Astronomical Union Symposium No 231*, 521–524 (2005).

11) Pilling S. and 8 co-authors. "Survival of gas-phase amino acids and nucleobases in space radiation conditions" in *Organic matter in space; Proceedings of the International Astronomical Union Symposium No 251*, 371–375 (2008).

Laboratory studies described in this paper predict that nucleobases – the building blocks of DNA and RNA – could survive the radiation environment of space even better than Amino Acids.

12) Rank D.M., Townes C.H. and Welch W.T. "Interstellar molecules and dense clouds". *Science* **174**, 1083–1101 (1971).

This review by Charles Townes and his team gives a comprehensive list of around 20 molecules that had been detected in the interstellar medium, including some whose identity was still in doubt in 1971. It is an early compendium of interstellar chemistry.

13) Salama F. "PAHs in astronomy: a review" in *Organic matter in space; Proceedings of the International Astronomical Union Symposium No 251*, 357–365 (2008).

Farid Salama proposes that polycyclic aromatic hydrocarbons are the building block of dust and that it will soon be possible to identify individual PAHs in the interstellar medium.

14) Schlemmer S., Asvany O., Hugo E. and Gerlich D. "Deuterium fractionation and ion-molecule reactions at low temperatures" in *Astrochemistry: Recent success and current challenges; Proceedings of the International Astronomical Union Symposium No 231*, 125–134 (2005).

It turns out that the abundance of Deuterium in interstellar molecules is even more complicated than simple considerations of the zero-point energy might imply.

15) Watson W.D. "The rate of formation of interstellar molecules by ion-molecule reactions". *Astrophysical Journal* **183**, L17–L20 (1973).

William Watson's paper is (one of) the first to argue that ion-molecule reactions involving H_3^+ are important for interstellar molecule formation as there is no energy barrier to such reactions occurring. Hence they can occur in low-temperature interstellar gas clouds.

16) Weaver H., Williams D.R.W., Dieter N.H. and Lum W.T. "Observations of a strong unidentified microwave line and of emission from the OH molecule". *Nature* **208**, 29–31.

Weaver and his team find that there is emission from the source W3 that they cannot identify: they associate it with "Mysterium", later found to be due to the OH molecule masing.

Chapter 5

1) For much of the background material in this book about Hawaii, three classic sources are extremely useful. They are:

1a) Carlquist S.J. *Hawaii: a natural history.* (American Museum of Natural History, Natural History Press, 1970).

Sherwin Carlquist's book gives enormous detail about the geology, the flora and the fauna of the Hawaiian chain, and how geology and geographical position gave rise to the unique character to the islands.

1b) Kamakau S.M. *The ruling chiefs of Hawaii.* (Revised edition: University of Hawai'i Press, 1994).

Samuel Manaiakalani Kamakau's book was written between 1866 and 1871 as a series of newspaper articles in the Hawaiian language newspapers, *Ke Au Oko a* and *Ka N pepa K oko a* before appearing as a book in 1961, first published by Kamehameha Schools Press. Kamakau was born in 1815, towards the end of the reign of Kamehameha I, and was able to discuss Hawaiian history with elders who had been alive before and at the time of Captain James Cook's arrival at the islands in 1778.

1c) Sahlins M. *How natives think ... about Captain Cook, for instance.* (2nd edition: University of Chicago Press, 1996).

Whether or not you agree with Marshall Sahlins' polemic against fellow anthropologist Gananath Obeyesekere about whether the Hawaiians who met Cook really believed him to be the god Lono, and the ramifications of that belief (or not), it gives huge insight into Hawai'i at the time of European contact.

2) Three websites were also very helpful in writing this chapter:

2a) http://pvs.kcc.hawaii.edu/

This is the website of the Polynesian Voyaging Society and contains a wealth of information about their past and current projects.

2b) http://mkooc.org/css-timeline.html

Andrew Pickles' website gives a concise timeline of the history of astronomy in Hawaii since World War II, which is then expanded in time and detail by ...

2c) http://www.ifa.hawaii.edu/users/steiger/introduction.html

... that of the pioneer of Hawaiian astronomy post-war, Walter Steiger. Steiger, sadly, died on February 6, 2011.

3) The following articles give some insight into the controversy over the use of Maunakea for astronomy:

3a) Chu R. and Ho N. "Should the Thirty-Meter Telescope project on Maunakea be allowed to proceed?" *Hawaii Business* **December 30** (2008).

In this pair of articles, Bank of Hawaii senior vice-president Chu and Hawaii Sierra Club Co-chair argue for and against the project to build the 30-Meter Telescope on Maunakea.

3b) McFarling U.L. "Science, culture clash over sacred mountain". *Los Angeles Times* **March 18** (2001).

This article looks carefully at both sides of the arguments, and the background to the Keck Observatory project to build outrigger telescopes.

3c) Pisciotta K. and Ho N. "Hawaiians and Sierra Club sue to reverse Keck outrigger decision". *Environmental Health Network* **November 30**, press release (2004).

This press release announces the intention of the Maunakea Ainana Hou and Sierra Club to contest the decision of the Hawaiian Board of Land and Natural Resources to allow the Keck outrigger project to proceed.

3d) "Hawaiian judge reverses permit for more Maunakea telescopes". *Environmental News Service* **August 7**, (2006).

This anonymous article carries news that a judge has upheld the Maunakea Ainana Hou and Sierra Club case against the building of outrigger telescopes, which stopped them being built.

Annotated References and Further Reading to Chapters 207

4) One of the most appropriate introductions to infrared spectroscopy for astronomical purposes is:

Tennyson J. *Astronomical spectroscopy: an introduction to the atomic and molecular physics of astronomical spectra* (Imperial College Press, London, 2005; 2nd edition 2010).

5) Cernicharo J. and 43 co-authors "The ISO/LWS far-infrared spectrum of IRC+10216". *Astronomy and Astrophysics* **315**, L201–L204 (1996).

The Infrared Space Observatory is reported to have discovered a large variety of Carbon-bearing molecules in this well evolved star, including C_3.

6) De Jong T., Clegg P.E., Soifer B.T., Rowan-Robinson M., Habing H.J., Houck J.R., Aumann H.H. and Raimond E. "IRAS observations of Shapley-Ames galaxies". *Astrophysical Journal* **278**, L67–L70 (1984).

This paper gave some early results of the first infrared survey of galaxies that were well characterized at optical wavelengths.

7) Kessler M.F. and 10 co-authors "The Infrared Space Observatory (ISO) mission". *Astronomy and Astrophysics* **315**, L27–L31 (1996).

This key paper describes the ISO mission, its capabilities and its expected outcome.

8) De Muizon M., Geballe T.R., d'Hendecourt L.B and Baas F. "New emission features in the infrared spectra of two IRAS galaxies". *Astrophysical Journal* **306**, L105–L108 (1986).

This paper argues that new features seen by ground-based observatories in the spectra of galaxies measured by IRAS are due to polycyclic aromatic hydrocarbons (PAHs), and candidate PAHs are discussed.

9) Neugebauer G. and 27 co-authors "The Infrared Astronomical Satellite (IRAS) mission". *Astrophysical Journal* **278**, L1–L6 (1984).

This key paper describes the IRAS mission, its capabilities and its expected outcome.

10) Sanders D.B. and Mirabel D.F. "CO detections and IRAS observations of bright radio spiral galaxies at cz ≤ 9000 kilometers per second". *Astrophysical Journal* **298**, L31–L35 (1985).

This paper linked ground-based measurements of CO to IRAS far-infrared galaxy measurements to show that active star-forming galaxies had more molecular gas in them.

11) Schutte W.A. and 10 co-authors "ISO-SWS observations of infrared absorption bands of the diffuse interstellar medium: the 6.2μm feature of aromatic compounds". *Astronomy and Astrophysics* **337**, 261–274 (1998).

ISO's first-time detection of the 6.2 micron feature linked to aromatic compounds, in evolved massive stars, is taken as proof of the existence of PAHs in these objects.

Chapter 6

1) There are many books that can be used to find out more about Jupiter. Three have been useful to me in this chapter:

1a) Bagenal F. and Dowling T. (eds.) *Jupiter: The Planet, Satellites, and Magnetosphere* (Cambridge University Press, 2005).

This book reviews Jupiter science in the era of the Galileo space mission, with a series of contributed chapters from leading planetary scientists.

1b) Rogers J.H. *The Giant Planet Jupiter* (Cambridge University Press, 1995).

John Rogers' "practical astronomy handbook" is a very good guide to amateur astronomers who want to understand what they when they observe the giant planet.

1c) Spencer J.R. and Mitton J. (eds.) *The great comet crash: the collision of Comet Shoemaker-Levy 9 and Jupiter* (Cambridge University Press, 1995).

Brought out rapidly in the wake of the impact of SL9 and Jupiter, the contributed chapters in this book give a good first look at the scientific results of the worldwide observing campaign.

Annotated References and Further Reading to Chapters 209

2) Other key books on Jupiter include:

2a) Harland D.M. *Jupiter Odessey: the story of NASA's Galileo mission.* (Springer Verlag, 2000).

2b) Meltzer M. *History of the Galileo mission to Jupiter.* (NASA History Office, 2004).

These two books give both historical and scientific overviews of the Galileo mission.

3) Sieff A. and 10 co-authors "Thermal structure of Jupiter's upper atmosphere derived from the Galileo probe". *Science* **376**, 102–104.

This paper gives the first science reports from the Galileo probe's entry into Jupiter's atmosphere.

4) Various authors "Special section on Comet Shoemaker-Levy 9". *Science* **267**, 1277–1392 (1995).

This special section of the journal of the American Association for the Advancement of Science contains six papers by various authors relating the first results from the worldwide observing campaign for the impact of Comet SL9 on Jupiter.

5) Various authors "Reports on Galileo Orbiter". *Science* **274**, 377–412 (1996).

This special section of *Science* gives the first reports of the Galileo orbiter's observations of the giant planet.

For the section on the Allende and other meteorites, I used:

6) Becker L. and Bunch T.E. "Fullerenes, fulleranes and polycyclic aromatic hydrocarbons in the Allende meteorite". *Meteoritics and Planetary Science* **32**, 479–487 (1997).

This paper outlines some of the fullerenes and PAHs found in the meteorite.

7) Clarke R.S. Jr., Jarosewich E., Mason B., Nelen J., Gomez M. and Hyde J.R. "The Allende, Mexico, meteorite shower". *Smithsonian contributions to the Earth sciences* **5** (1971).

This is a comprehensive review of the early science surrounding the meteorite, from gathering the pieces through to chemical analysis.

8) Connolly J.N., Amelin Y., Krot A.N. and Bizzarro M. "Chronology of the Solar System's oldest objects". *Astrophysical Journal* **675**, L121–L124 (2008).

This paper dates fragments of the Allende meteorite to 4,565 million years.

9) Hsu W., Guan Y., Leshin L.A. and Ushikubo T. "Late episode of irradiation in the early Solar System: evidence from extinct ^{36}Cl and ^{26}Al in meteorites". *Astrophysical Journal* **640**, 525–529 (2006).

Looking at isotope ratios in meteorite samples, this paper argues for a burst of excess solar radiation during the early stages of the Solar System.

10) Wark D.A. "Birth of the presolar nebula: the sequence of condensation revealed in the Allende meteorite". *Astrophysics and Space Science* **65**, 275–295 (1979).

This paper argues that analysis of the Allende meteorite supports the theory that a supernova shock to a cloud of interstellar gas and dust triggered the formation of the Solar System, and that some of the meteorite's carbonaceous inclusions are "pre-solar".

For the sections on comets, amongst other references, I used:

11a) Crovisier J. "The molecular composition of comets and its interrelation with other small bodies of the Solar System". in *Asteroids, Comets, Meteors; Proceedings of the International Astronomical Union Symposium No 229*, 133–152 (2005).

11b) Schulz R. "Compositional coma investigations: gas and dust production rates in comets". in *Asteroids, comets, meteors; Proceedings of the International Astronomical Union Symposium No 229*, 413–423 (2005).

Between them, these two papers from the same IAU symposium give a good overview of the chemical composition of comets.

12) In these three papers, Fred Whipple develops his "dirty snowball" model of cometary nuclei, although he does not actually use that term.

Annotated References and Further Reading to Chapters 211

12a) Whipple F.L. "A comet model I: the acceleration of Comet Enke". *Astrophysical Journal* **111**, 375–394 (1950).

12b) Whipple F.L. "A comet model II: physical relations for comets and meteors". *Astrophysical Journal* **113**, 464–474 (1951).

12c) Whipple F.L. "A comet model III: the zodiacal light". *Astrophysical Journal* **121**, 750–771 (1955).

For the Deep Impact results, *Science* once more brought out a special issue:

13) Various authors "Deep Impact". *Science* **310**, 257–286 (2005).

Several multi-authored papers here (one with over 100 authors) testify to the huge interest in NASA's comet collision mission, and to the results that scientists were able to derive from it in a very short space of time.

14) Stardust results were also reported first in *Science*, again as a series of multi-authored papers reports the results of analyzing the material returned by the Stardust mission from Comet Wild 2, with overviews from Joanne Baker, Mike A'Hearn and Don Burnett:

Various authors "Special Section: Stardust". *Science* **314**, 1707–1739 (2006).

Interlude 2

For this Interlude, I have chosen the following papers arranged in the historical order with which they cover key moments in the discovery of H_3^+ in space.

1) Phillips T.G., Blake G.A., Keene J., Woods R.C. and Churchwell E. "Interstellar H_3^+: possible detection of the 1_{10}–1_{11} transition of H_2D^+". *Astrophysical Journal* **294**, L45–L48 (1985).

Tom Phillips' team made their tentative detection of the deuterated form of H_3^+ in NGC2264, but not in the Taurus Molecular Cloud. This got a weak "maybe" from the interstellar community.

2) Three papers covered the discovery of H_3^+ in Jupiter, Saturn and Uranus:

2a) Drossart P. and 11 co-authors "Detection of H_3^+ on Jupiter". Nature **340**, 539–541 (1989).

2b) Geballe T.R., Jagod M.-F. and Oka T. "Detection of H_3^+ infrared emission lines in Saturn". Astrophysical Journal **410**, L109–L112 (1993).

2c) Trafton L.M., Geballe T.R., Miller S. and Tennyson J. "Detection of H_3^+ from Uranus". Astrophysical Journal **405**, 761–766 (1993).

3) Meyer W., Botschwina P. and Burton P. "*Ab initio* calculation of near-equilibrium potential and multipole moment surfaces and vibrational frequencies of H_3^+ and its isotopomers". Journal of Chemical Physics **84**, 891–900 (1986).

This was the first calculation of the potential energy and dipole moment surfaces of H_3^+ that was sufficiently accurate for the infrared spectrum to be calculated from first principles and used for chemical and astronomical purposes.

4) Two papers from Jonathan Tennyson's group, using the Meyer, Botschwina and Burton potential, calculated the exact spectrum from which the H_3^+ lines in Jupiter were first identified:

4a) Miller S. and Tennyson J. "Calculated rotational and ro-vibrational transitions in the spectrum of H_3^+, Astrophysical Journal **335**, 486–494 (1988).

4b) Miller S. and Tennyson J. "Hot band transitions in H_3^+: first principles calculations". Journal of Molecular Spectroscopy **136**, 223–240 (1989).

5) Geballe T.R. and Oka T. "Detection of H_3^+ in interstellar space". Nature **384**, 334–335 (1996).

At last, H_3^+ is detected in the interstellar medium, where it had been predicted.

Chapter 7

1) The area of exoplanet research is expanding so rapidly, with more claims, counter-claims and counter-counter-claims being made that it is nearly impossible to keep up with the literature. Jean Schneider's keeps people updated with the number and type of exoplanets that have been found at http://exoplanet.eu/.

These are a few of the key papers that map out how our story has taken shape during the last quarter of a century.

2) Campbell B, Walker G.A.H. and Yang S. "A search for sub-stellar companions to solar-type stars". *Astrophysical Journal* **331**, 902–911 (1988).

Bruce Campbell and his colleagues found seven stars that showed signs of having companions with masses between 1 and 10 Jupiter masses.

3) Koskinen T.T., Aylward A.D. and Miller S. "A stability limit for the atmospheres of giant extrasolar planets". *Nature* **450**, 845–848 (2007).

Tommi Koskinen's team show that H_3^+ cooling is key to the stability of the atmospheres of giant exoplanets as they migrate closer to their parent star.

4) Lissauer J.J. and 38 co-authors "A closely packed system of low-mass, low-density planets transiting Kepler-11". *Nature* **470**, 53–58.

As one of the early results from NASA's Kepler mission, the discovery of this complex planetary system promises much more exciting exoplanet science to come.

5) Mayor M. and Queloz D. "A Jupiter-mass companion to a solar-type star". *Nature* **378**, 355–359 (1995).

This is the paper that opened up the "modern" age of extrasolar planets, detecting a Jupiter-like planet around the star 51 Pegasi.

6) Ribas I., Guinan E.F., Güdel M. and Audard M. "Evolution of the solar activity over time and effects on planetary atmospheres: I high energy irradiances (1-1700Å)". *Astrophysical Journal* **622**, 680–694 (2005).

This paper shows that our Sun gave out much more ionizing radiation in its early years than at present.

7) Tinetti G. and 12 co-authors "Water vapour in the atmosphere of a transiting extrasolar planet". *Nature* **448**, 169–171 (2007).

In this paper, Giovanna Tinetti analyse data from NASA's Spitzer orbiting telescope to show that there is water vapour in the atmosphere of exoplanet HD 189733b.

8) Wolszczan A. and Frail D.A. "A planetary system around the millisecond pulsar PSR1257+12". *Nature* **355**, 145–147 (1992).

Most astronomers recognize this paper as being the first truly believable paper that reported an extrasolar planetary system.

Chapter 8

1) The scientific material that this chapter deals with is to be found in many journals, ranging from planetary science through to astrobiology. A textbook that covers much of the material in this book is:

Shaw A.M. *Astrochemistry: from astronomy to astrobiology.* (John Wiley and Sons, Chichester, 2006).

For the sections on prebiotic molecules, and their relationship to life, useful references include:

2) Blank J.G., Miller G.H., Ahrens M.J. and Winans R.E. "Experimental shock chemistry of aqueous amino acid solutions and the cometary delivery of pre-biotic compounds". " *Origins of Life and Evolution of the Biosphere* **31**, 15–51 (2001).

Based on laboratory experiments, this paper argues that comets could have delivered pre-formed Amino Acids to Earth.

Annotated References and Further Reading to Chapters 215

3) Beltran M.T., Codella C., Viti S., Neri R. and Cesaroni R. "The first detection of Glycoaldehyde outside the Galactic Center". *Astrophysical Journal* **690**, L93-L96 (2009).

This paper shows that simple sugars can be produced on interstellar grains within the first 100,000 years of a stellar system forming.

4) Dartnell L. *Life in the universe: a beginner's guide.* (One world Publications, Oxford, 2007).

This book takes over where the present one leaves off, looking at how life might develop on Earth and elsewhere. Lewis Dartnell also explores the possibilities of life existing outside of Earth in: Dartnell L. "Biological constraints on habitability". *Astronomy and Geophysics* **52**, 1.25–1.28 (2011).

5) Davies P.C.W., Benner S.A., Cleland C.E., Lineweaver C.H., McKay C.P. and Wolfe-Simon F. "Signatures of a shadow biosphere". *Astrobiology* **9**, 241–249 (2009).

In this paper, Paul Davies' team argue that if life forms readily in conditions such as those that prevailed on the early Earth, it may have formed here several times. If so, then scientists might look for signs of shadow, Life 2. The paper discusses how this goal might be pursued.

6) Jenniskens P. and 34 co-authors "The impact and recovery of asteroid 2008 TC_3". *Nature* **458**, 485–488 (2009).

This paper reported on the recovery and initial analysis of 2008 TC_3. This was followed up by a special issue of the journal *Meteoritics and Planetary Science* (**45**, 1553–1845 (2010)) devoted to analysis of the meteorite.

7) Krasnopolski V.A., Maillard J.-P. and Owen T.C. "Detection of methane in the martian atmosphere: evidence for life?". *Icarus* **172**, 537–547 (2004).

This paper explored the possibility of a biological origin for Martian Methane discovered, originally, by the European Space Agency's Mars Express mission. It was followed later by claims from NASA's Mars Climate Orbiter team that they had also detected Methane of biological origin.

8) Miller S.L. "The production of amino acids under possible primitive Earth conditions". *Science* **117**, 528–529 (1953).

This paper inspired the famous "Miller-Urey experiment", which was reported in:

9) Miller S.L. and Urey H. C. "Organic compound synthesis on the primitive Earth". *Science* **130**, 245–251 (1959).

10) Sagan C. and Khare B.N. "Tholins: organic chemistry of interstellar grains and gas". *Nature* **277**, 102–107 (1979).

Sagan and Khare argue that very complex organic molecules may form in the interstellar medium, but would not have survived the formation of Earth.

11) Stano P. and Luisi P.L. "Basic questions about the origins of life: proceedings of the Erice International School on Complexity (fourth course)". *Origins of Life and Evolution of the Biosphere* **37**, 303–307 (2007).

This review of the Erice conference posed key questions about what was necessary for life to have evolved, and how to understand that process.

12) Tolstogulov V. "Why were polysaccharides necessary?" *Origins of Life and Evolution of the Biosphere* **34**, 571–597 (2004).

The answer, according to this paper, is because they helped to concentrate pre-biotic molecules so that they could form biotic aggregates.

13) There are regular conferences on astrobiology throughout the world. In the writing of this book, the Eighth European Workshop on Astrobiology that was held in 2008 in Neuchatel, Switzerland had its presentations reported in:

Various authors "Special Issue" *Origins of Life and Evolution of the Biosphere* **39**, 1–89 (2009).

14) Watson J.D. and Crick F.H.C. "A structure for Deoxyribose Nucleic Acid". *Nature* **171**, 737–738 (1953).

The paper that announced the Double Helix!

15) Wolfe-Simon F. and 12 co-authors "A bacterium that can grow by using Arsenic instead of Phosphorus". *ScienceExpress* **2 December**, 1–7 (2010).

This paper reported the existence of a bacterium from Lake Mono in California that substituted Arsenic for Phosphorus in key biolmolecules, including its DNA.

16) Two papers from 1996 in *Science* look at claims to have found fossilized bacteria in the martian meteorite ALH 84001.

16a) McKay D.S. and 8 co-authors "Search for past life on Mars: possible relic biogenic activity in martian meteorite ALH 84001". *Science* **273**, 924–930 (1996).

This is the announcement from David McKay that his team's analysis of ALH 84001 shows up structures that look like microfossils. Richard Kerr provides commentary in:

16b) Kerr R.A. "Ancient life on Mars?". *Science* **273**, 864–866 (1996).

Titan science is covered by:

17) Lebreton J.-P. and 11 co-authors "An overview of the descent and landing of the Huygens probe on Titan". *Nature* **438**, 758–764 (2005).

This paper outlines the short life of this remarkable probe on Titan.

18) Lorenz R. and Mitton J. *Lifting Titan's veil: exploring the giant moon of Saturn.* (Cambridge University Press, 2002).

This book gives a very accessible overview of scientists' understanding of Titan prior to the arrival of Cassini/Huygens.

19) Norman L.H. and Fortes A.D. "Is there life on … Titan?" *Astronomy and Geophysics* **52**, 1.39 – 1.42 (2011).

Given the chemical and physical conditions prevailing on Titan, these authors argue that it meets the conditions for life to form.

20) Sims I. (chair) "Chemistry of the Planets". *Faraday Discussions* **147** (2010).

Although this conference at St Jacut de la Mer was advertised as discussing the chemistry of planetary atmospheres in general, many of the scientific papers given focused on the chemistry of Titan's atmosphere, derived both from measurements from Huygens and Cassini, and from laboratory simulations and measurements, as well as chemical models.

21) Stofan E.R. and 37 co-authors "The lakes of Titan". *Nature* **445**, 61–64 (2007).

After several years of searching, Cassini at last produced evidence of Hydrocarbon lakes on Titan.

22) Tobie G., Lunine J.I. and Sotin C. "Episodic outgassing as the origin of Methane on Titan. *Nature* **440**, 61–64 (2006).

Jonathan Lunine's team here proposed cryovolcanism and outgassing as a source of Methane on Titan, given that the lakes did not seem large enough to explain it wholly.

23) Waite J.H. Jr., Young D.T. Cravens T.E., Coates A.J., Crary F.J., Magee B. and Westlake J. "The process of Tholin formation in Titan's upper atmosphere". *Science* **316**, 870–875 (2007).

Results from Cassini reported here showed that enormous organic molecules were forming in the upper atmosphere of Titan.

Epilogue

1) In two key papers, Alan Carrington and his team reported the spectrum of the H_3^+ molecular ion at the moment that it was breaking up. The first reported the spectrum they recorded, the second gave more details of this and showed that at "pseudo low resolution" it had a number of discrete peaks that indicated symmetry was being maintained.

1a) Carrington A. Buttenshaw J. and Kennedy R. "Observation of the infrared spectrum of H_3^+ at its dissociation limit". *Molecular Physics* **45**, 753–758 (1982).

1b) Carrington A. and Kennedy R.A. "Infrared predissociation spectrum of the H_3^+ ion". *Journal of Chemical Physics* **81**, 91–112 (1984).

2) Miller S. and Tennyson J. "Calculation of the high angular momentum dissociation limit for H_3+ and H_2D+". *Chemical Physics Letters* **145**, 117–120 (1988).

This paper looked to see at what point H_3^+ would break up if it were spun up to rotate very energetically.

3) Brass O., Tennyson J. and Pollak E. "Spectroscopy and dynamics of the highly excited non-rotating 3D H_3^+ molecular ion". *Journal of Chemical Physics* **92**, 3377–3386 (1990).

This paper gives the first convincing demonstration that the breakup of H_3^+ as a result of large vibrational energies might give rise to the spectrum measured by Alan Carrington's team.

Some Useful Numbers

In working out how much mass is contained in stars, planetary atmospheres and the gas of the Interstellar Medium, some numbers may be helpful:

The mass of a Hydrogen atom, $m_H = 1.6735 \times 10^{-27}$ kg (kg = kilogram)

The mass of the proton, $m_p = 1.6726 \times 10^{-27}$ kg

The mass of the neutron, $m_n = 1.6749 \times 10^{-27}$ kg

The mass of the electron, $m_e = 0.00091 \times 10^{-27}$ kg

Avogadro's Number is the number of atoms or molecules in a kilogram atomic or molecular weight of a substance, $A_N = 6.022 \times 10^{26}$ atoms/molecules per kg A/M Wt

1 m³ (cubic metre) of Water weights 10^3 kg = 1 tonne

1 m³ of Water contains 3.343×10^{28} molecules

1 m³ of air at the International Union of Pure and Applied Chemistry Standard Temperature (273.15K, 0°C) and Pressure (0.986 atmosphere) (STP) contains 2.652×10^{25} molecules

1 m³ of air at STP weighs 1.269 kg

The Interstellar Medium (ISM), as a mixture of n(H) Hydrogen atoms and n(H_2) Hydrogen molecules per cubic metre (m⁻³) "weighs" (if it were in Earth's gravity) n(H) × 1.6735 x 10^{-27} kg + n(H_2) × 3.3470 × 10^{-27} kg

The typical Hot ISM contains pure H 10^6 m⁻³ and "weighs" 1.6735×10^{-21} kg m⁻³

The typical ypical Diffuse ISM contains 50/50 H/H_2 10^8 m⁻³ and "weighs" 2.5103×10^{-19} kg m⁻³

Some Useful Numbers

The typical Dense ISM contains 10%/90% H/H_2 10^{10} m^{-3} and "weighs" 3.1797×10^{-17} kg m^{-3}

The typical Core ISM contains pure H_2 10^{12} m^{-3} and "weighs" 3.3470×10^{-15} kg m^{-3}

The ionization rate of the ISM by cosmic rays, $\varsigma = 10^{-17} s^{-1}$ for the Dense ISM

$\varsigma = 10^{-16} s^{-1}$ (or more) for the Diffuse ISM

Pictures and Figures

1. Takeshi Oka at work in his laboratory at the University of Chicago: *credit – Oka Ion Factory, University of Chicago*
2. J.J. Thomson giving a lecture demonstration in the Cavendish Laboratory at the University of Cambridge: *credit – The Cavendish Laboratory, University of Cambridge*
3. The energy levels of the Hydrogen Atom: a robin on an Xmas Tree can jump all the way to the lowest branch or hop down branch-by-branch, giving up much less energy per hop: *credit – Steve Miller*
4. Alex Dalgarno: *credit – Harvard University, Institute for Theoretical Atomic, Molecular and Optical Physics*
5. Jonathan Tennyson: *credit – University College London, Department of Physics and Astronomy*
6. Gerhard Herzberg: *credit – Canadian Astronomical Society*
7. A view of our Milky Way galaxy, showing "lanes" due to gas and dust clouds: *credit – NASA*
8. The Orion Nebula shows trails due to "bullets" of rapidly moving gas passing through it: *credit – Gemini Observatory*
9. Mary Lea Heger and her husband Donald Shane: *credit – Donald E. Osterbrock, courtesy AIP Emilio Segre Visual Archives, Physics Today Collection*
10. In the Interstellar Medium, H_3^+ initiates a web of reactions that lead on to the formation of larger molecules. This "web" involves chemistry with Carbon- and Oxygen-containing molecules: *credit – Steve Miller*
11. Carbon forms isomers, molecules with the same overall chemical composition, but with different arrangements that have different properties. On the left is Acetaldehyde. The starred Hydrogen atom in the top left moves onto the Oxygen in the molecule on the

224 Pictures and Figures

right to form Vinyl Alcohol. (Key: Carbon atom – black; Oxygen atom – red; Hydrogen atom – yellow.): *credit – Steve Miller*

12. sp^3 Carbon forms *optical* isomers, molecules with the same overall chemical composition, but with different arrangements. The central starred Carbon atom is surrounded by four different groups – a Hydrogen atom, a Methyl group, an OH group (Alcohol), and a NH$_2$ group (Amine). The left and right-hand molecules are mirror images of one another. (Key: Carbon atom – black; Oxygen atom – red; Nitrogen atom – blue; Hydrogen atom – yellow.): *credit – Steve Miller*

13. The summit of Maunakea with the Maunakea Observatory: *credit – Richard Wainscoat, Institute for Astronomy, Hawaii*

14. An artist's impression of the Infrared Astronomical Satellite in orbit: *credit – NASA*

15. An artist's impression of the James Webb Space Telescope in orbit: *credit – NASA*

16. The vibrational energy levels of a diatomic molecule may be represented by a potential energy "valley" with "bridges" crossing it at various levels: *credit – Steve Miller*

17. The rotational energy levels of a diatomic molecule may be considered as a system of "ladders" fixed to the vibrational "bridges". Note that the rotational ladder on one bridge can go past the bridge above it, all the way to the top of the potential energy "valley": *credit – Steve Miller*

18. An artist's impression of Galileo probe entering the atmosphere of Jupiter. In the background the orbiting spacecraft looks on: *credit – NASA*

19. A fragment of the Allende meteorite held at the Natural History Museum in London: *credit – Natural History Museum, London*

20. The launch of ESA's Rosetta mission on March 2, 2004: *credit – the European Space Agency*

21. An infrared image of Jupiter after the impacts of Comet Shoemaker-Levy 9. A fragment is seen hitting on the dawn (left) limb of the planet. Elsewhere, impact scars from previous impacts can be seen: *credit – the Max Planck Institute, Germany*

22. An artist's impression of the highly energetic young star W33A, with its dark dusty accretion disk and powerful jets of gas pouring out of the top and bottom: *credit – Gemini Observatory, artwork by Lynette Cook*

23. The Arecibo Radio Observatory in Puerto Rico: *credit – NAIC – Arecibo Observatory, a facility of the NSF*
24. The Huygen's probe's view of its landing site on Titan: *credit – the European Space Agency*
25. NASA has constructed this image of Sotra Facula a cryovolcano on the surface of Titan, based on data returned by the radio science instruments on Cassini: *credit – NASA*

Index

A
Acetic acid, 70
A'Hearn, Mike, 130
Alanine, 172
Aldebaran, 34, 182
Aldehyde, 79–81, 89
Algae, 117
ALH84001, 186, 187
Alkali metal, 5
Alkanes, 78, 79, 84
Alkenes, 79, 84
Alkynes, 79, 82
Allende meteorite, 124, 128, 186
Alpha particles/rays (a particle/rays), 14, 28, 33
Aluminum, 26, 27, 88, 125, 129
American Association for the Advancement of Science (AAAS), 70
Amino acid, 79, 84–86, 89, 125, 126, 131, 172, 173, 183, 186, 188
Ammonia (NH_3), 81, 86, 90, 120, 127, 129, 137, 171, 172, 178
Andromeda galaxy/constellation, 103, 157
Anglo-Australian Telescope, 38
Anharmonicity, 111
Anion, 2, 3, 5, 28, 44, 48, 52, 81, 86, 182, 183, 185
Anode, 2, 45, 48, 50

Antares, 34
Anthracene, 86, 87
Anthropic principle, 32
Aquila, constellation, 65, 165
Arcturus, Hoku Lea, 34, 93, 182
Arecibo Observatory, 158
Argon, 26–28, 61, 79, 91, 178
Argonne National Laboratory, 58
Arianne 5
Aristotle, 7
Arpanet (Advanced Research Project Agency network), 144
Arsenic (As), 189
Asteroid, 115, 123, 124, 126, 133, 186
Aston, Francis, 47
Astronomical and Astrophysical Society of America, 93
Astrophysical Data System, 103
Atmospheric Physics Laboratory, University of Michigan, 146
Atom, 2, 4–6, 10–24, 27–31, 34, 35, 37, 38, 41, 45, 46, 48–60, 63, 65, 66, 68, 69, 72–89, 96, 104, 105, 107–112, 121, 125, 126, 140, 142, 146, 147, 151, 167, 168, 182–184
Aurora/aurorae (Borealis, Australis), 7, 40, 122, 123, 137, 138, 141, 142, 143
Aylward, Alan, 167, 168

227

B

Bacteria, 187
Barium (Ba), 125
Beagle, Beagle 2, 174, 176
Becklin, Eric, 97, 148
Bell Laboratories, 60
Benzene (C_6H_6), 85, 86, 121, 180–182, 184
Beryllium, 26, 31, 32, 48, 125
Berzelius, Jöns Jacob, 171
Beta particles/rays (b particles/rays), 14
Betelgeuse, 34, 37, 65
Bielefeld University, 112
Big Bang, 5, 10–13, 19, 21, 31, 32, 43, 104, 189
Black hole, 6, 42, 43, 89, 150
Black smoker, 185
Bohr, Niels (Bohr atom), 16, 17, 19, 27, 50–53, 105
Born-Oppenheimer approximation, 54, 105
Boron, 26, 27, 32
Botschwina, Peter, 112
Boyle, Robert *(The Skeptical Chymist)*, 26
de Broglie, Louis, 53
Bromine, 49
Brown Dwarf, 156, 157, 161
Brownian motion, 13
Brownlee, John,
Buckminster Fuller, Richard, 87
Burbidge, Margaret/Geoffrey, 33
Burns, John, 94
Burton, Peter, 112

C

C_{60}, 4, 125
Calcium (Ca), 27, 38, 125, 129, 137
California Institute of Technology, 33, 147
California Sub-millimeter Observatory (CSO), 97
Campbell, Bruce, 156, 157, 162
Canada France Hawaii Telescope (CFHT), 97, 138, 142, 155, 156
Carbon–14, 30, 173
Carbon (C), 4, 6, 10, 11, 26–28, 30, 32–35, 38, 41, 44, 48–50, 59, 67, 72–74, 77–88, 104, 119–121, 125–127, 129, 130, 136, 175, 182–185, 189
Carbonaceous chondrite, 125, 128, 133
Carbon dioxide (CO_2), 28, 48, 77, 90, 127, 131, 132, 153, 154, 171–173, 181, 183
Carbon monoxide (CO), 38, 39, 41, 43, 70, 72, 78, 88, 97, 103, 104, 127, 129, 131, 149
Carbon sulfide (CS), 38
Carney, G.D., 111
Carrington, Alan,
Cassini, Giovanni, 138
Cassini-Huygens mission, 176, 178
Cassiopeia A gas cloud, 69
Cathode, 2, 45, 47, 48
Cathode ray tube, 45
Cation, 2–6, 16, 27, 41, 44, 49–52, 58, 60, 61, 77, 86, 116, 140, 168, 181, 182
Cat's Eye nebula, 41
Cavendish Laboratory, Cambridge University, 3, 45
21-Centimeter radiation, 68
Central Molecular Zone, 150
Centre Européenne de Recherche Nucléaire/European Nuclear Research Center (CERN), 9, 58
Centre Spatial Guyanais, 100
Ceres, 124
Cesium, 28
Chadwick, James, 29, 51
Chain reaction, 89, 168
Charbonneau, David, 166
Chemical elements, 10, 11, 13, 15, 26, 27, 31–33, 43, 46, 49, 50, 64, 118

Chernobyl, 28
Chicago, University of, 1–3, 5, 143, 172
Chirality, 82, 83, 183, 188
Chlorine (Cl), 2, 4, 5, 26–30, 49, 50, 125
Christmas Tree Cluster, 148
Christofferson, Ralph, 58
Chromium (Cr), 129, 137
Chyba, Chris, 175
Classical mechanics, 13, 75, 89
Clinton, President Bill, 187
Cobalt (Co), 38
Columbia University, 51
Comet Churyumov-Gerasimenko, 131
Comet Shoemaker-Levy–9, 134, 136
Comet Tempel–1, 130, 131
Comet Wild–2, 129, 130
Conroy, Harold, 58
Cook, Captain James, 93, 96
Cosmic Microwave Background Radiation (CMBE), 12, 19, 43
Cosmic ray ionization rate, 151
Coulomb explosion, 59
Crick, Francis, 183
Crookes, William (Crookes tube), 45, 46
Curie, Marie, 29
Curl, Bob, 87
Cyanic acid, 171
Cyanobacteria, 117
Cyanogen (C_2N_2), 180
Cyclo-propane, butane, pentane, hexane, 84, 85
Cygnus constellation, Cygnus OB2–12, 150
Cytosine, 184, 188

D
Dalgarno, Alex, 39–42
Dalton, John, 13
Dark matter, 19
Darwin, Charles, 174

DCN, 76
Deep Impact mission, 130
Democritus, 13
Dempster, A.J., 49, 71
Dense Interstellar Medium, 65
Deoxyribonucleic acid (DNA), 20, 85, 183, 184, 186, 189
Deuterium (D), 11, 20, 21, 24, 30–33, 35, 51, 52, 56, 74, 76, 104, 107, 147, 157
Diffuse Interstellar Bands (DIB), 67, 87
Diffuse Interstellar Medium, 65, 76
Dirac, Paul, 75
Doppler shift, 148, 156, 157
Double bond, 79–82, 85
Dust, 6, 13, 36, 60, 63–67, 88, 89, 102, 103, 115, 116, 125, 128–131, 134, 135, 137, 142, 149, 158
Dwarf planet, 124

E
Earth, 4, 6, 7, 19, 25, 31, 36, 42, 43, 54, 64, 65, 68, 84, 91, 92, 94, 96, 99–101, 104, 115, 117–119, 121, 122, 124, 126–131, 133, 134, 136, 138–142, 153–156, 158–166, 168, 169, 172, 175–179, 181–189
Eddington, Arthur, 162, 163
Einstein, Albert, 13, 16, 28, 52, 162, 163
Electron (e^-), 2, 3, 5, 9–12, 14–22, 27–30, 34, 35, 42, 44–48, 50, 51, 53–59, 65, 66, 71, 72, 76–78, 81, 85, 86, 89, 104, 109, 111, 116, 140, 142, 150, 186
Electronic orbital, 58
Electron volt, 9, 76
Enceladus, 176
Energy barrier, 71–74, 82, 89
Enzyme, 185, 187

Ester, 81
Ethane (C_2H_6), 78, 79, 84, 121, 131, 154, 178, 180
Ethyl formate, 150
Europa, 154, 175, 176
Europa Jupiter System mission, 154, 175
European Space Agency (ESA), 36, 100, 103, 128, 131, 132, 154, 174, 176, 177
Europlanet network, 165
Extrasolar Planets Encylcopaedia, 162
Extrasolar planets/exoplanets, 162, 164
Eyring, Henry, 52, 53, 56, 57

F

Faraday, Michael, 1, 45, 46
Fluorapatite, 185
Fluorine, 26, 27, 49
Formaldehyde (H_2CO), 70, 80, 81, 131, 185
Fourier transform spectrometer/spectrometry, 138, 142
Fowler, Willie, 33
Fractionation effect, 74
Frail, Dale, 159, 160, 162
Franklin, Rosalind, 183
von Fraunhofer, Joseph, 15
French Guiana, Kourou, 99
Fullerene, 87, 125
Fundamental band (vibration)/line, 143, 144

G

Galaxy, 3, 6, 7, 19, 25, 36, 40, 43, 62–65, 74, 89, 103, 149, 150, 155
Galileo spacecraft/mission, 115, 120, 175
Gallium, 27
Gamma Cephei/Cepheus constellation, 157
Gamma particles/rays (g particles/rays), 14, 43, 96
Gamow, George, 10, 11, 31
Ganymede, 154, 175, 176
Gas Analysis Package (GAP), 174
Gas clouds, 6, 19–21, 23, 24, 32, 33, 36–39, 62, 64, 65, 67, 69–71, 73, 76, 88, 103, 104, 115, 117, 125, 161
Geballe, Tom, 146–151
Geiger, Hans, 14
Gemini Telescope, 97
Germanium (Ge), 27
GFAJ-1, 189
Giotto di Bondone, 128
Giotto mission, 129
GL2136, 149, 150
Glycine, 172
Glycolaldehyde, 186
Goldilocks zone, 164
Gould, Gordon, 60
Grain, 13, 88–90, 104, 121, 122, 125–127, 129, 130, 133, 149, 185
Great Red Spot, 118
Great Salt Lake, Utah, 129
Guanine, 184

H

H_3^+, 4–7, 21, 24, 35, 40–44, 49–53, 55–62, 64, 71–73, 76, 77, 82, 85, 86, 107, 108, 110–113, 116, 122, 123, 137, 143–151, 153, 166–169, 171, 181
Haldane, J.B.S., 10, 172
Haleakala, 92–94, 99
Hale Pohaku, 138, 145
Halley's Comet, 93, 127–129, 134
Halogen, 5
H-alpha (Hα), 17, 66
Harvard Smithsonian Center for Astrophysics, 103
Hat Creek Observatory, 69
Hawaii, 34, 40, 53, 91–99, 130, 138, 142, 148, 155, 156, 177, 181

Hawaiian Astronomical Society, 93
Hawaii State Board of Land and Natural Resources, 98
Hawking, Stephen, 42
H_2D^+, 76, 116, 148
HD 40307/b, 164
HD 189733/b, 165, 166
HD 209458/b (Osiris), 166
HDO, 76
Heavy bombardment, 126, 185
Heger, Mary Lea, 67, 68, 87
Heisenberg, Werner (Heisenberg Uncertainty Principle), 74
Helium (He), 4, 6, 10, 11, 14, 15, 20, 21, 26–35, 38, 43, 44, 77, 79, 102, 103, 117–119, 121, 122, 140, 157, 165
Herbst, Eric, 72
Hertz, Heinrich, 28
Herzberg, Gerhard, 59, 156
Herzberg Institute, 59–62, 143–145, 148, 155, 156
Hilo, 53, 97, 98, 148
Hirschfelder, Joseph, 52, 53, 56–58, 60, 61, 108, 111
Hogness, T.R., 49, 50, 71
Horsehead Nebula, 66, 67
Hot band (vibrational), 143, 145
Hoyle, Fred, 32–34
H_3^+ thermostat, 167–169
Hubble Space Telescope (HST), 36, 100, 130, 134, 165
Hunter Waite, Jack, 182
Huygens, Christiaan, 176
Hydrogen
  anion H⁻, 44, 48
  atom H, 2, 4, 5, 10, 11, 16–21, 23, 24, 28–30, 37, 41, 45, 50, 51, 53, 55, 56, 65, 66, 68, 72–75, 78–81, 83, 84, 86, 88, 89, 105, 107–109, 111, 121, 146, 148, 167, 168
  molecule H_2, 5, 21, 23, 24, 42, 50, 51, 56, 70–73, 75, 76, 79, 88, 105, 107–109, 167, 168

Hydrogenated amorphous carbon (HAC), 88
Hydrogen cyanide (HCN), 74, 76, 82, 86, 103, 104, 131, 137, 178, 180, 181, 184
Hydrogen fluoride (HF), 104, 156

I

Icy core, 119
Illinois, University of, 72
Impact, 80, 130–133, 135–137, 186
Impact crater, 175, 178
InfraRed Astronomical Satellite (IRAS), 102–104
Infrared astronomy, 96, 97, 99, 100, 138
Infrared Space Observatory (ISO), 103, 104
Institut d. Astrophysique de Paris (IAP), 138, 165
Institut de Physique Nucléaire, Lyon, 58
Institute for Astronomy, Hawaii, 95, 97, 177
International Astronomical Union (IAU)/IAU circular, 37, 134
International Ultraviolet Explorer (IUE), 140, 142
Interstellar medium (ISM), 62, 64, 65, 67, 70, 73, 74, 76, 77, 85–88, 101, 115, 121, 125, 129, 130, 132, 133, 147–151, 171, 182, 184–186
Io, 123, 142
Ion, 1–3, 5, 76, 113, 148
Ion Factory, 1–3, 5, 113, 148
IRC +10216, 77, 88, 104
Iron (Fe), 15, 35, 38, 125, 128, 129, 137
Isomers, 79–81, 83, 84
Isotope, 28, 30, 31, 33, 34, 51, 91, 126, 147, 179

J

James Clark Maxwell Telescope (JCMT), 97
James Webb Space Telescope (JWST), 100–102, 104, 165
JANET. *See* Joint Academic Network (JANET)
Jeffries, John, 95, 96
Jenniskens, Peter, 186
Jet Propulsion Lab, Pasadena, 120, 131
Jodrell Bank Radio Observatory, 67–68, 70
Joint Academic Network (JANET), 144
Joliot-Curie, Irene, 29
Joliot, Frederic, 29
Jupiter, 6, 40, 69, 104, 115–124, 131, 133–138, 140–147, 154, 156–162, 164–169, 175, 176, 185

K

Kaimuki Observatory, 93
Kalakaua, King, 93
Keck Observatory, 98
Ken's House of Pancakes, 97
Kepler mission, 164
Ketone, 79
Khartoum University, 186
Kitt Peak National Observatory, 150
Klemperer, Bill, 72
Koskinen, Tommi, 168
Kourou, French Guiana, 99, 100
Kroto, Harry, 87
Kuiper Airborne Observatory, 101, 147
Kuiper, Gerard, 94, 95, 101

L

Labeled Release experiment, 173
Lake Mono, 189
Large Hadron Collider (LHC), 9, 12
Large Magellanic Cloud, 7, 36, 40
Las Campanas Observatory, Chile, 25
Laser, 60, 61, 69
Leiden Institute of Chemistry, 126
Lenard, Philipp, 28
Leo, constellation, 77
Lepp, Stephen, 39, 41, 42
Leucippus, 13
Levy, David, 134
Lick Observatory, 67, 161
Life, 3, 4, 7, 10, 20, 25–62, 72, 86, 93, 96, 111, 115, 117, 120, 122, 126, 130, 153–155, 160, 165, 166, 169, 171–189
Light curve, 37, 38
Lithium, 11, 20, 21, 26, 31, 35, 36
Local Bubble, 65, 140
Lovell, Sir Bernard, 68, 69
Lucretius, *De Rerum Natura*, 13, 14
Luisi, Pier Luigi, 187, 188
Lunar and Planetary Institute, 94
Lunine, Jonathan, 179, 180
Lunn, E.G., 49, 50, 71
Lyndon B. Johnson Space Center, 186

M

Magnesium, 26, 38, 125, 129, 137
Magnetic field, 45, 70, 116, 119, 139, 141, 175
Magnetosphere, 139–141, 176, 181
Magnetotail, 141
Maillard, Jean-Pierre, 138, 142–145
Maiman, Theodore, 60
Main Belt, 124
Marcy, Geoff, 161
Mars, 34, 124, 126, 133, 138, 153, 154, 161, 162, 166, 172–176, 186, 187
Marsden, Ernest, 14
Martins, Zita, 126
Maser, 69, 70

Mass over charge ratio, 48, 49
Maui, 34, 92, 93, 99
Maunakea, 91, 92, 94–99, 101, 130, 131, 138, 142, 148, 155
Maunakea Anaina Hou, 99
Mayor, Michel, 160–162, 164
McDaniel, Earl, 71, 147
McKay, David, 186, 187
Meikle, Peter, 38–42
Mendeleev, Dimitri, 27
Mercury, 25, 48, 133, 160–163, 168, 176, 182
Methane (CH_4), 49, 78, 82–85, 104, 120, 121, 127, 129, 131, 137, 138, 154, 166, 172, 177–181, 183
Methanol (CH_3OH), 70, 104, 131
Meyer, Wilfred, 112
Milky Way, 3, 6, 19, 63–65, 69, 101, 103, 104, 150, 151
Miller, Stanley, 172
Millikan, Robert, 48
2.72 Millimeter line, 70,
Millisecond Pulsar, 157–159
Mitchell, Joni, 129
Molecular rotation, 22, 57, 62, 104, 110
Molecular vibration, 57, 105, 107, 109
Molecule, 2–7, 20–24, 35, 38, 39, 41–43, 46, 48–59, 61–63, 67, 69–113, 117, 119, 121, 122, 125, 126, 128, 130–132, 137, 140, 142–151, 165–168, 172, 173, 175, 178, 181–185, 187–189
Monoceros, constellation, 87, 148
Moon, 104, 122, 124, 126, 141, 154, 163, 171, 175–179, 181, 183
Moore, Patrick, 135
Mount Palomar Observatory, 94
Mount Wilson Observatory, 94
Murchison meteorite, 125, 126, 128, 186
Mysterium, 70

N
Nanotube, 87
NASA Infrared Telescope Facility (IRTF), 96–97, 123, 148
National Aeronautical and Space Agency (NASA), 36, 64, 94–97, 100–102, 115, 116, 120, 123, 129, 130, 135, 140, 147, 148, 154, 164, 165, 174–176, 180, 186, 189
National Radio Astronomy Observatory, 159
Neodymium, 125
Neon, 26–28, 34, 47, 79, 119, 121
Neptune, 104, 117, 140, 146, 161
Neugebauer, Gerry, 97
Neutrino, 25
Neutron (n), 2, 9, 11, 29–31, 33–35, 51, 147, 157, 159, 160
Neutron star, 34, 35, 157, 159, 160
New Galactic Catalogue, 41, 148
Newton, Sir Isaac *(Opticks)*, 13, 15, 26, 155
New York State University, 111
Nickel (Ni), 129
Nitrogen (N), 4, 6, 10, 11, 26, 27, 32, 34, 35, 41, 66, 72–74, 79, 82, 83, 86, 100, 119, 120, 126, 136, 140, 153, 172, 177, 178, 180, 183, 184
Nobel Prize, 59, 69, 87, 183
Noble gases, 79
Nucleobases, 85, 126, 184, 186

O
Observatoire de Paris, 138
OH maser, 70
Oka, Takeshi, 1, 2, 59, 111, 143
Olivine, 92, 104, 125, 129
Onizuka, Ellison (Onizuka Center), 98
Opacity, stellar, 44
Oparin, Alexander, 172

Open University, 174
70 Ophiuchi, 155
Opik-Oort Cloud, 127
Orion constellation/nebula/
　Molecular Cloud (OMC),
　65–67, 69, 148
Overtone band (vibrational)/line,
　145
Owen, Toby, 177, 182
Oxygen (O), 2, 4, 6, 10, 11,
　26–28, 31, 32, 34, 35,
　38, 41, 48, 50, 70, 72, 73,
　77–81, 83, 86, 88, 95, 100,
　117, 119–121, 136, 140,
　171, 172, 177, 178, 181,
　183, 184
Ozone (O_3), 4, 140

P

Pauling, Linus, 183
51 Pegasi, 160–162, 164
Penzias, Arno, 12
Peptide, 173
Periodic table, 27, 49
Permafrost, Siberian, 175
Phillips, Tom, 147
Phoenix lander, 175
Phosphates, 183–185
Phosphorus (P), 26, 27, 120, 183,
　184, 189
Photon, 9, 12, 16–20, 22, 23, 25,
　28, 29, 35, 43, 44, 63, 76, 88,
　89, 121, 142
Pillinger, Colin, 174
Pioneer mission (Pioneer 10, Pioneer 11), 141
Planck constant (h), 16, 75
Planck, Max, 15, 136
Planetary nebula, 34, 41, 115
Plasma, 7, 65, 140, 141, 175, 181
Pluto, 124
Polarity, 81
Pollak, Eli,
Polonium, 29

Poly-cyclic aromatic hydrocarbons
　(PAH), 86, 87, 103, 125, 130,
　131, 182, 186
Polymerization, 81
Porter, R.N., 111
Positive rays, 48, 52
Potassium, 38, 49, 91
Potential energy surface, 54, 55,
　75, 105, 106, 109, 111, 112
Pre-dissociation,
Propane (C_3H_8), 78, 80, 84
Propyl cyanide, 150
Protein, 79, 86, 131, 172, 183, 188
Proto-galaxy, 19, 35, 43
Proton (p^+), 2, 5, 6, 9–11, 16–18, 21,
　28–31, 33–35, 41, 44, 51, 54,
　58, 59, 65, 71, 72, 85, 147
Proton affinity, 72
Proto-solar nebula, 115, 117–120,
　122, 124–127, 130, 132, 175
PSR B1257$_+$, 159, 160
Punahou, 93
Purine, 126, 184
Pyrene, 87
Pyrimidine, 126, 184
Pyroxene, 125

Q

Quantum/quantum mechanics,
　16, 27, 51, 53–57, 61, 74, 75,
　82, 85, 89, 105, 107, 111, 112,
　119, 145
Queloz, Didier, 160–162

R

Radical, 67, 78, 80, 85, 86, 121
Radio astronomer/astronomy, 12,
　67, 69, 71, 72, 93, 147
Recombination Era, 12, 19–21
Red Giant, 34, 65, 154, 182
Red Rectangle, 87, 88, 104
Relativity, special/general, 13
Resonance, 85

Rhodamine, 61
Ribonucleic acid (RNA), 85, 126, 183–186, 188, 189
Ribose, 183–186
Rice University, Texas, 87
Rigel, 65, 102
Rosetta mission, 131, 132
Royal Astronomical Society, 37, 39, 88
Royal Institution, 1, 45–47
Royal Order of Kamehameha I, 99
Royal Society, 26, 47, 148
Rubidium, 49
Rutherford, Ernest, 14–16, 27–29, 46
Rydberg constant, 16, 17, 19

S
Sagan, Carl, 126, 182
Sagittarius, constellation, 65
Saturn, 6, 104, 117, 138, 140, 141, 146, 154, 158, 161, 166, 176, 178, 179, 181
Scandium, 27
Schawlow, Arthur, 60
Scorpio constellation, 93
Search for Extra-Terrestrial Intelligence (SETI), 159
SETI Institute, 175, 186
Shaddad, Muawia, 186
Shelton, Ian, 25, 37
Shoemaker, Margaret/Gene,
Sierra Club, 99
Silicon (Si), 26, 27, 35, 38, 88, 103
Silicon carbide (SiC), 103
Silicon oxide (SiO), 38
Single bond, 82, 85
Smalley, Rich, 87
Solar system, 6, 7, 35, 38, 54, 64, 65, 101, 104, 115, 117, 122, 124–128, 130, 132–135, 137, 140, 141, 145, 146, 153–155, 160–162, 165–167, 169, 175, 176, 178, 182, 183, 186

Solar wind, 7, 139, 141, 142, 145, 177, 181
Southampton, University of,
Spectroscopy, 15, 48, 59, 112, 113, 138, 143, 145
Spitzer mission, 130, 165
Sputnik, 69
Stano, Pasquale, 187, 188
Stardust mission, 129, 130
Steiger, Walter, 93
Sugar, 81, 171, 183, 185, 186
Sulfur, 25–27, 35, 38, 120, 125, 133, 142, 181
Sulfuric acid, 153
Summer triangle, 33
Sun, 6, 7, 15, 19, 20, 25, 32–36, 41, 42, 44, 54, 64, 92, 93, 99, 100, 103, 115–122, 125, 127, 129–132, 134, 138–141, 154–158, 160–169, 176, 181, 182
Sun in Time project, 169
Supergiant star, 34–37, 43
Supernova/Supernova 2006gy/ Supernova 1987A, 25, 34–39, 41–43, 64, 65, 70, 71, 101, 115, 125, 129, 137, 150, 157–159
Sussex, University of, 87
Sutcliffe, Brian, 111
Symmetry, 53, 107–109, 113, 147, 167

T
Tarantula, 1, 5
Taurus constellation/Molecular Cloud (TMC), 33, 65, 67, 76, 148, 163
2008 TC_3, 186
Tennyson, Jonathan, 40, 111, 112, 165
Thatcher, Margaret, 128
Three Gorges power station, China, 141
Thymine, 184

Tinetti, Giovanna, 165
Titan, 104, 154, 176–184
Townes, Charles, 60, 69, 70, 72
Trafton, Larry, 145
Triple alpha process, 33, 34
Triple bond, 79
Tritium (T), 31, 51
T-Tauri phase/star, 33
Tunneling, 9, 87, 89, 135

U
Ultraviolet heating, 168
Ultraviolet radiation (UV), 4, 35, 37, 38, 132, 169
United Kingdom Infrared Telescope (UKIRT), 97, 131, 145, 146, 148–150
University College London (UCL), 39, 40, 140, 143, 144, 167
University of Hawaii 88? Telescope, 95, 96
Uracil, 126, 184, 186, 188
Uranium, 14, 28, 35
Uranus, 6, 104, 117, 140, 146, 161, 164
Urea, 171, 172
Urey, Harold, 51, 172

V
Valency, 46, 50, 52
Vega spacecraft, 128
Vela, constellation, 65
Venus, transit of, 93
Vibrational band/level, 61
Vidal-Madjar, Alfred, 166
Viking spacecraft (Viking 1, Viking 2), 173
Vinyl alcohol, 81
Vinyl chloride, 81
Volcano, cryo-volcano, 91, 92, 117, 142, 176, 180, 181

Voyager mission (Voyager 1, Voyager 2), 140, 142, 145, 182
Vulpecula constellation, 165

W
W33A, 149–150
Washington, University of, 129
Water, 2–4, 6, 7, 14, 20, 25, 26, 43, 69, 70, 72, 76, 86, 88, 90, 91, 96, 99, 104, 111, 117, 118, 120, 122, 127–129, 131–133, 136, 143, 150, 154, 161, 165, 171–177, 179, 181, 183, 184
Watson, Bill, 72
Watson, James, 183
Watson, Jim, 62, 111, 143–144
Wave function, 53, 89
Wave-particle duality, 53
Weaver, Harold, 69, 70
Weizmann Institute, Tel Aviv, 59, 144
Welles, Orson, 173
Wells, H.G., 173
Whipple, Fred, 127, 128
White Dwarf, 34, 77
Wilkins, Maurice, 183
Williams, David, 88
Wilson, Robert, 12
Wilson, Sir Robert (Bob), 140
Wisconsin, University of, 56
Wöhler, Friedrich, 171
Wolfe-Simon, Felisa, 189
Wolszcznan, Aleksander, 159, 160, 162

X
Xanthine, 126

Z
Zero Point Energy, 74, 76, 147